RAPID INFECTIOUS DISEASES AND TROPICAL MEDICINE

Rachel Isba
Oxford University Medical School,
The John Radcliffe Hospital,
Oxford

EDITORIAL ADVISOR
Brian J. Angus
Clinical Tutor in Medicine,
Honorary Consultant Physician,
Nuffield Department of Medicine,
University of Oxford,
The John Radcliffe Hospital,
Oxford

SERIES EDITOR
Amir Sam
Royal Free and University College Medical School,
University College London,
London

Blackwell
Publishing

For my Mum

Blackwell Publishing, Inc., 350 Main Street, Malden, Massachusetts 02148-5020, USA
Blackwell Publishing Ltd, 9600 Garsington Road, Oxford OX4 2DQ, UK
Blackwell Publishing Asia Pty Ltd, 550 Swanston Street, Carlton, Victoria 3053, Australia

First published 2004

Library of Congress Cataloging-in-Publication Data

Isba, Rachel.
Rapid infectious diseases and tropical medicine / Rachel Isba; editorial advisor, Brian Angus—1st ed.
p.; cm.—(Rapid series)
Includes bibliographical references.
ISBN 1-4051-1325-1
1. Communicable diseases—Handbooks, manuals, etc. 2. Tropical medicine—Handbooks, manuals, etc.
[DNLM: 1. Communicable Diseases—diagnosis—Handbooks. 2. Communicable Diseases—therapy—Handbooks. 3. Signs and Symptoms—Handbooks. 4. Tropical Medicine—methods—Handbooks. WC 39 176r 2003] I. Title. II. Series.

RC112.I83 2003
616.9—dc22 2003019589

ISBN 1-4051-1325-1

A catalogue record for this title is available from the British Library

Set in 7.5/9.5 pt Frutiger
by Kolam Information Services Pvt. Ltd, Pondicherry, India

Commissioning Editor: Vicki Noyes
Editorial Assistant: Nicola Ulyatt
Production Editor: Jonathan Rowley
Production Controller: Kate Charman

For further information on Blackwell Publishing, visit our website: http://www.blackwellpublishing.com.

CONTENTS

Contents

Foreword

Rapid Infectious Diseases and Tropical Medicine has been written by an Oxford University clinical medical student partly while she was travelling in the South Atlantic but mainly as part of her special study module on medical publishing. I can think of no better way to study medical publishing than by publishing a book!

There are many reasons that nowadays the student of general medicine needs a book like this. The growth in rapid worldwide travel has meant that many previously geographically obscure infections are now rapidly at our door-step and in our clinics. The spectre of bioterrorism has meant that we now need to be vigilant to unusual and exotic infection and the growth of the multi-drug resistant organisms within a hospital environment increasingly populated with immunocompromised patients has meant that rapid identification and control of infection is essential. Even the tabloid press now regularly feature articles about MRSA, the superbug!

This book aims to allow the rapid identification of the key features of infectious diseases organised in a simple and easily accessible way. It also should help clarify communication between the laboratory and the ward as it is organised in such a way that either the organism itself or the disease it causes can be searched for.

This will be a valuable resource for undergraduates revising for final BM as well as postgraduates revising for MRCP and MRCPath. Although common in clinical practise, infection is not a usual clinical scenario in exams for obvious reasons but tends to be well represented in written papers. We hope that you will find this book useful.

Brian Angus

Ab	Antibody
ABPA	Allergic BronchoPulmonary Aspergillosis
ACh	AcetylCholine
AFB	Acid-Fast Bacilli
Ag	Antigen
AIDS	Acquired Immune Deficiency Syndrome
ALT	ALanine Transaminase
ANS	Autonomic Nervous System
APTT	Activated Partial Thromboplastin Time
ARDS	Adult Respiratory Distress Syndrome
ARF	Acute Renal Failure
AST	ASpartate Transaminase
ATLL	Adult T-cell Leukaemia/Lymphoma
ATN	Acute Tubular Necrosis
AXR	Abdominal X-Ray
BAL	BronchoAlveolar Lavage
BCG	Bacille Calmette–Guérin
C.	Central
CAH	Chronic Active Hepatitis
cAMP	cyclic Adenosine MonoPhosphate
CAPD	Continuous Ambulatory Peritoneal Dialysis
CD4+	Cluster of Differentiation 4 positive
CF	Cystic Fibrosis
CF	Complement Fixation
CFS	Chronic Fatigue Syndrome
CHF	Chronic Heart Failure
CHIK	CHIKungunya
CIN	Cervical Intraepithelial Neoplasia
CJD	Creutzfeldt–Jakob Disease
CK	Creatine Kinase
CMI	Cell-Mediated Immunity
CMV	CytoMegaloVirus
CNS	Central Nervous System
COPD	Chronic Obstructive Pulmonary Disease
CPH	Chronic Persistent Hepatitis
CRP	C-Reactive Protein
CSF	CerebroSpinal Fluid
CT	Computerised Tomography
CTF	Colorado Tick Fever
CV	CardioVascular
CXR	Chest X-Ray
DEN	DENgue
DF	Dengue Fever
DHF	Dengue Haemorrhagic Fever
DIC	Disseminated Intravascular Coagulation
DNA	DeoxyriboNucleic Acid
DOTS	Directly Observed Treatment Short course
DSS	Dengue Shock Syndrome
DTP	Diphtheria Tetanus Pertussis (vaccine)
DTwP	Diphtheria Tetanus whole-cell Pertussis (vaccine)
EBV	Epstein–Barr Virus
ECG	ElectroCardioGram
EEE	Eastern Equine Encephalitis
EEG	ElectroEncephaloGram
ELISA	Enzyme-Linked ImmunoSorbent Assay
EM	Electron Microscope
ENL	Erythema Nodosum Leprosum
ENT	Ear Nose Throat
EPI	Expanded Programme of Immunisation
ERCP	Endoscopic Retrograde Cholangio Pancreatography
ESR	Erythrocyte Sedimentation Rate
ETBE	European Tick-Borne Encephalitis
FBC	Full Blood Count
FTA	Fluorescent Treponemal Antibody
G +ve	Gram stain positive
G −ve	Gram stain negative
GBS	Guillain–Barré Syndrome
GH	Growth Hormone
GIT	GastroIntestinal Tract
GU	GenitoUrinary

List of Abbreviations

H	Haemagglutinin
H & E	Haematoxylin & Eosin
HACEK	Haemophilus Actinobacillus Cardiobacterium Eikenella Kingella
HAV	Hepatitis A Virus
Hb	Haemoglobin
HB	Hepatitis B (vaccine)
HBV	Hepatitis B Virus
HBVeAg	Hepatitis B Virus envelope Antigen
HBVsAg	Hepatitis B Virus surface Antigen
HCV	Hepatitis C Virus
HDV	Hepatitis D Virus
HEV	Hepatitis E Virus
HHV	Human Herpes Virus
Hib	Haemophilus influenzae b (vaccine)
HIV	Human Immunodeficiency Virus
HLA-B27	Human Lymphocyte Antigen B27
HPV	Human Papilloma Virus
HSV	Herpes Simplex Virus
HTLV	Human T-cell Leukaemia Virus
HUS	Haemolytic Uraemic Syndrome
HVB	Herpes Virus B
HZV	Herpes Zoster Virus
IBS	Irritable Bowel Syndrome
ID	IntraDermal
IFA	ImmunoFluorescence Assay
IFN	InterFeroN
Ig	Immunoglobulin
IM	IntraMuscular
IP	Incubation Period
IPV	Inactivated Polio Vaccine
ITU	Intensive Treatment Unit
IU	International Units
IUCD	IntraUterine Contraceptive Device
IV	IntraVenous
IVDA	IntraVenous Drug Abuse
JBE	Japanese B Encephalitis
JCV	Jamestown Canyon virus
LCMV	Lymphocytic ChorioMeningitis Virus
LDH	Lactate DeHydrogenase
LFT	Liver Function Tests
LRTI	Lower Respiratory Tract Infection
mAb	monoclonal Antibody
MACELISA	IgM Antibody Capture ELISA
MALToma	Mucosa-Associated Lymphoid Tissue-oma
MCV	Mean Cell Volume
MDRTB	MultiDrug Resistant TuBerculosis
MMR	Measles Mumps Rubella (vaccine)
MRI	Magnetic Resonance Imaging
MRSA	Methycillin-Resistant Staphylococcus Aureus
MSU	MidStream Urine
N	Neuraminidase
N.	North
Na	Sodium
NGU	Non-Gonococcal Urethritis
nvCJD	new variant Creutzfeldt–Jakob Diseases
N.W.	North West
N.W.	New World
OCP	Oral Contraceptive Pill
OPV	Oral Polio Vaccine
OT	Occupational Therapy
O.W.	Old World
PCR	Polymerase Chain Reaction
PEP	Post-Exposure Prophylaxis
PID	Pelvic Inflammatory Disease
PGL	Persistent Generalised Lymphadenopathy
PMN	PolyMorphoNucleocyte
PO	Per Os
PPI	Proton Pump Inhibitor
PUD	Peptic Ulcer Disease
PV	Per Vaginum
RBC	Red Blood Count
RhF	Rheumatic Fever
RIF	Right Iliac Fossa

RMSF	Rocky Mountain Spotted Fever
RNA	RiboNucleic Acid
RSSE	Russian Spring–Summer Encephalitis
RSV	Respiratory Syncitial Virus
RT	Reverse Transcriptase
°S	degrees South
S.	South
SARS	Severe Acute Respiratory Syndrome
SC	SubCutaneous
SCID	Severe Combined ImmunoDeficiency
SDH	SubDural Haematoma
S.E.	South East
SIADH	Syndrome of Inappropriate AntiDiuretic Hormone secretion
SF	Scarlet Fever
SOB	Short Of Breath
SOL	Space-Occupying Lesion
spp.	species
SRSV	Small Round Structured Virus
SSA	Sub-Saharan Africa
SSSS	Staphylococcal Scalded Skin Syndrome
STSS	Streptococcal Toxic Shock Syndrome
STI	Sexually Transmitted Infection
TB	TuBerculosis
TFP	Tropical Flaccid Paralysis
TPHA	Treponema Pallidum HaemAgglutination
TSP	Tropical Spastic Paraparesis
TSS	Toxic Shock Syndrome
TT	Tetanus Toxoid (vaccine)
U & E	Urea & Electrolytes
URTI	Upper Respiratory Tract Infection
USS	UltraSound Scan
UTI	Urinary Tract Infection
VDRL	Venereal Disease Research Laboratory
VEE	Venezuelan Equine Encephalitis
VHF	Viral Haemorrhagic Fever
ViCPS	typhoid Purified PolySaccharide (vaccine)
VZV	Varicella Zoster Virus
W.	West
WBC	White Blood Count
WEE	Western Equine Encephalitis
WNV	West Nile Virus
XR	X-Ray
YF	Yellow Fever
Z–N	Ziehl–Neelsen
/24	hours
/7	days
/52	weeks
/12	months
1°	primary
2°	secondary
3°	tertiary
4°	quaternary
♀	female
♂	male
↑	increased
↓	decreased
→	goes to
↔	goes both ways
/	or
$>$	greater than
$<$	less than
$\geq$	greater than or equal to
$\leq$	less than or equal to
$\gg$	much greater than
$\ll$	much less than

Rapid Series Mnemonic

D:	Definition	*Doctors*
A:	Aetiology	*Are*
A/R:	Associations/Risk factors	*Always*
E:	Epidemiology	*Emphasising*
H:	History	*History-taking &*
E:	Examination	*Examining*
P:	Pathology	*Patients*
I:	Investigations	*In*
M:	Management	*Managing*
C:	Complications	*Clinical*
P:	Prognosis	*Problems*

PART 1: SIGNS & SYMPTOMS

Fever

Viruses
Dengue
EBV
Hepatitis (prodromal)
HIV
Influenza
Viral haemorrhagic fevers (Ebola, Lassa, Marburg, etc.)
Yellow fever

Bacteria
Borrelia sp.
Brucella sp.
Coxiella burnetii (Q fever)
Francisella tularensis (tularaemia)
Legionella pneumophila
Salmonella paratyphi
Salmonella typhi
Streptococcus spp.
Yersinia pestis

Mycobacteria
Mycobacterium tuberculosis

Fungi
Coccidioides immitis
Histoplasma spp.

Protozoa
Entamoeba histolytica
Leishmania spp.
Plasmodium spp.
Trypanosoma spp.

Helminths
Hyperinfection syndromes

Spirochaetes
Leptospira interrogans (Weil's disease)

Other organisms
Rickettsia spp.

Sepsis

Bacteria
Almost any, but particularly
Enterobacter spp.
Enterococcus spp.
Escherichia coli
Klebsiella spp.
Listeria monocytogenes
Neisseria meningitidis
Pseudomonas aeruginosa
Salmonella spp.
Staphylococcus aureus

Streptococcus faecalis
Streptococcus pneumoniae
Streptococcus pyogenes A–T

Mycobacteria
Mycobacterium tuberculosis

Fungi
Candida spp.

Cardiovascular

Endocarditis

Native valve
Bacteria
HACEK
Staphylococcus aureus
Streptococcus viridans

Other organisms
Chlamydia spp.
Coxiella burnetii
Mycoplasma spp.

Prosthetic valve
Bacteria
Coliform
Enterococcus spp.
Staphylococcus aureus
Staphylococcus epidermidis

Fungi
Candida spp.

IV drug abusers
Bacteria
Coliform
Enterococcus faecalis
Pseudomonas aeruginosa
Staphylococcus aureus

Fungi
Candida spp.

Myocarditis

Viruses
Adenovirus
CMV
Coxsackie A & B
EBV
Echovirus
HIV
Influenza
Mumps

Bacteria
Borrelia (Lyme disease)
Corynebacterium diphtheriae
Neisseria meningitidis
Staphylococcus aureus

Mycobacteria
Mycobacterium tuberculosis

Fungi
Candida spp.

Protozoa
Toxoplasma gondii
Trypanosoma cruzi

Helminths
Trichinosis

Other organisms
Coxiella burnetii
Chlamydia psittaci

Pericarditis
Viruses
Adenovirus
Coxsackie
Echovirus
EBV
Influenza
Mumps

Bacteria
Staphylococcus aureus
Streptococcus pneumoniae
Streptococcus pyogenes

Mycobacteria
Mycobacterium tuberculosis

Fungi
Histoplasma

Protozoa
Entamoeba histolytica

Upper respiratory tract/ENT

Coryza
Viruses
Coronavirus
Rhinovirus

Croup (acute laryngotracheobronchitis)
Viruses
Adenovirus

Influenza
Measles
Parainfluenza 1, 2 or 3
Rhinovirus
RSV

Bacteria
Corynebacterium diphtheriae
Haemophilus influenzae

Other organisms
Mycoplasma spp.

Epiglottitis
Viruses
Varicella zoster

Bacteria
Haemophilus influenzae b
Haemophilus influenzae (non-b)
Staphylococcus aureus
Streptococcus pneumoniae
Streptococcus pyogenes

Oral infection
Viruses
Herpes simplex
Varicella zoster

Fungi
Candida albicans

Otitis externa
Bacteria
Pseudomonas aeruginosa
Staphylococcus aureus

Otitis media
Bacteria
Haemophilus influenzae
Moraxella catarrhalis
Pseudomonas aeruginosa
Staphylococcus aureus
Streptococcus pneumoniae
Streptococcus pyogenes

Mycobacteria
Mycobacterium tuberculosis

Other organisms
Chlamydia trachomatis
Mycoplasma spp.

Pharyngitis/tonsillitis

Viruses
Adenovirus
CMV
Coxsackie
EBV
Herpes simplex
HIV
Parainfluenza

Bacteria
Corynebacterium spp.
Neisseria gonorrhoeae
Neisseria meningitidis
Streptococcus group C & G β-haemolytic
Streptococcus pyogenes

Protozoa
Toxoplasma gondii

Other organisms
Mycoplasma spp.

Sinusitis

Acute
Bacteria
Haemophilus influenzae b
Moraxella catarrhalis
Staphylococcus aureus
Streptococcus pneumoniae
Streptococcus pyogenes

Chronic
Bacteria
Haemophilus influenzae b
Moraxella catarrhalis
Pseudomonas aeruginosa
Staphylococcus aureus
Streptococcus milleri
Streptococcus pneumoniae
Streptococcus pyogenes

Mycobacteria
Mycobacterium tuberculosis

Fungi
Aspergillus spp.
Mucormycosis

Lower respiratory tract

Bronchiolitis

Viruses
Parainfluenza
RSV

Bronchitis

Bacteria
Haemophilus influenzae
Pseudomonas aeruginosa
Staphylococcus aureus
Streptococcus pneumoniae

Cystic fibrosis (infections in)

Bacteria
Pseudomonas aeruginosa
Staphylococcus aureus

Empyema

Bacteria
Actinomyces spp.
Clostridium welchii
Streptococcus faecalis
Streptococcus pneumoniae

Protozoa
Entamoeba histolytica

Higher organisms
Nocardia spp.

Lung abscess

Bacteria
Actinomyces spp.
Klebsiella pneumoniae
Staphylococcus aureus
Streptococcus faecalis
Streptococcus milleri

Protozoa
Entamoeba histolytica

Higher organisms
Nocardia spp.

Pneumonia

Community-acquired

Viruses
Influenza

Bacteria
Haemophilus influenzae b
Legionella pneumophila
Mycoplasma spp.
Staphylococcus aureus
Streptococcus pneumoniae

Other organisms
Chlamydia spp.

Nosocomial
Bacteria
Enterobacteria spp.
Klebsiella spp.
Legionella pneumophila
Pseudomonas aeruginosa
Staphylococcus aureus

Aspiration
Bacteria
Anaerobes from oropharynx

Compromised host
Bacteria
Haemophilus influenzae
Legionella pneumophila
Moraxella catarrhalis
Other Gram negatives
Staphylococcus aureus
Streptococcus pneumoniae

Fungi
Cryptococcus neoformans
Histoplasma capsulatum
Pneumocystis carinii

Other organisms
Mycoplasma spp.

Gastrointestinal

Colitis

Bacteria
Clostridium difficile

Enteric fever

Bacteria
Salmonella paratyphi A, B & C
Salmonella typhi

Food-borne disease

Viruses
Calicivirus
HAV
SRSV

Bacteria
Bacillus cereus
Campylobacter jejuni
Clostridium botulinum
Clostridium perfringens
Escherichia coli
Salmonella spp.
Staphylococcus aureus
Vibrio cholerae

Toxins
Ciguatera fish poisoning
Diarrhoeic shellfish poisoning
Paralytic shellfish poisoning
Scrombotoxin fish poisoning

Nausea, vomiting and diarrhoea
Viruses
Calicivirus
Rotavirus
SRSV

Bacteria
Bacillus cereus
Campylobacter jejuni
Clostridium spp.
Escherichia coli
Salmonella spp.
Shigella spp.
Vibrio cholerae
Yersinia spp.

Mycobacteria
Mycobacterium tuberculosis

Protozoa
Cryptosporidium
Entamoeba histolytica
Giardia lamblia
Isospora belli
Microspora

Helminths
Schistosoma spp.
Strongyloides

Oesophagitis
Viruses
Herpes simplex

Fungi
Candida spp.

Peritonitis
Bacteria
Enterobacter spp.
Enterococcus spp.
Escherichia coli
Streptococcus pneumoniae

Mycobacteria
Mycobacterium tuberculosis

Fungi
Cryptococcus neoformans

Tropical sprue, enteropathy

'Malabsorption in deprived areas of the tropics where no bacterial, viral or parasitic infection can be detected.'

Whipple's disease

Bacteria
Tropheryma whippelii

Hepatitis

Acute viral

Adenovirus
CMV
EBV
Enterovirus
HAV
HBV
HBV & HDV coinfection
HEV
HSV
Yellow fever

Chronic viral

HAV (extremely rarely)
HBV
HBV & HDV coinfection
HCV

Urinary tract infection

Catheter-associated/complicated/renal abscess

Bacteria
Enterobacteriaceae
Enterococci
Escherichia coli
Proteus
Pseudomonas aeruginosa
Staphylococcus aureus

Uncomplicated

Bacteria
Escherichia coli
Klebsiella
Proteus
Pseudomonas aeruginosa
Staphylococcus saprophyticus

Other organisms
Chlamydia trachomatis
Mycoplasma spp.

Haematuria

Viruses
Adenovirus

Bacteria
Escherichia coli
Neisseria gonorrhoeae

Helminths
Schistosoma haematobium

Sterile pyuria
Mycobacteria
Mycobacterium tuberculosis

Genitourinary

Epididimitis
Bacteria
Enterobacter
Neisseria gonorrhoeae

Helminths
Filariasis

Other organisms
Chlamydia trachomatis

Orchitis
Viruses
Coxsackie B
Mumps

Bacteria
Neisseria gonorrhoeae

Helminths
Filariasis

Other organisms
Chlamydia trachomatis

Prostatitis
Bacteria
Escherichia coli
Enterococcus
Staphylococcus aureus

Sexually transmitted infections
Viruses
HBV
HCV
HDV
HIV 1 & 2
HPV
HSV
HTLV 1 & 2

Bacteria
Neisseria gonorrhoeae
Haemophilus ducreyi

Fungi
Candida albicans

Spirochaetes
Treponema pallidum (syphilis)

Other organisms
Chlamydia trachomatis

Urethritis
Bacteria
Neisseria gonorrhoeae

Other organisms
Chlamydia trachomatis

Vulvovaginitis
Bacteria
Neisseria gonorrhoeae
Streptococcus pyogenes
Trichomonas vaginalis

Fungi
Candida albicans

Central nervous system
Brain abscess
Bacteria
Bacteroides fragilis
Burkholderia pseudomallei
Enterobacteria
Staphylococcus aureus
Streptococcus milleri
Streptococcus pneumoniae

Mycobacteria
Mycobacterium avium
Mycobacterium tuberculosis

Protozoa
Toxoplasma gondii

Encephalitis
Viruses
BKV
CMV
EEE
EBV
HIV 1
HSV 1 & 2
JBE
JCV
Mumps
Polio
Rabies
Rubella
VZV
WEE

Meningitis

Acute

Viruses
Adenovirus
Coxsackie
Echovirus
Enteroviruses
Herpes simplex
HIV
Mumps
Polio
Varicella zoster

Bacteria
Escherichia coli
Haemophilus influenzae
Listeria monocytogenes
Neisseria meningitidis
Staphylococcus aureus
Streptococcus pneumoniae

Mycobacteria
Mycobacterium tuberculosis

Fungi
Cryptococcus neoformans

Chronic

Mycobacteria
Mycobacterium avium
Mycobacterium tuberculosis

Fungi
Cryptococcus neoformans

Spirochaetes
Treponema pallidum (syphilis)

Neuritis

Viruses
EBV
CMV
Influenza
Polio

Eyes

Conjunctivitis

Viruses
Adenovirus
Enterovirus
HSV
Measles
Rubella
VZV

Bacteria
Haemophilus influenzae
Neisseria gonorrhoeae
Neisseria meningitidis
Staphylococcus aureus
Streptococcus pneumoniae

Mycobacteria
Mycobacterium tuberculosis

Spirochaetes
Leptospira
Treponema pallidum

Endophthalmitis

Bacteria
Staphylococcus aureus
Staphylococcus epidermidis
Streptococcus spp.

Keratitis

Viruses
Adenovirus
HSV
HZV
Measles
Mumps

Bacteria
Neisseria gonorrhoeae
Pseudomonas aeruginosa
Staphylococcus aureus
Streptococcus pneumoniae
Streptococcus pyogenes

Mycobacteria
Mycobacterium tuberculosis

Fungi
Candida spp.

Spirochaetes
Treponema pallidum (syphilis)

Periocular

Bacteria
Haemophilus spp.
Staphylococcus aureus
Streptococcus pyogenes

Skin & soft tissue infection

Cellulitis

Normal host

Bacteria
Clostridium welchii

Staphylococcus aureus
Streptococcus pyogenes

Diabetic
As above plus

Bacteria
Anaerobes
Enterococcus
Coliform

Lymphadenopathy

General
Viruses
CMV
EBV
HAV
HBV
HIV
HSV
Measles
Rubella
VZV

Bacteria
Bartonella
Borrelia (Lyme disease)
Brucella spp.
Francisella tularensis (tularaemia)
Salmonella paratyphi
Salmonella typhi
Yersinia pestis

Mycobacteria
Mycobacterium tuberculosis

Fungi
Coccidioides
Histoplasma

Protozoa
Leishmania spp.
Toxoplasma gondii
Trypanosoma spp.

Spirochaetes
Treponema pallidum (syphilis)

Other organisms
Rickettsia typhi

Granulomatous
Mycobacteria
Mycobacterium tuberculosis

Fungi
Histoplasma

Protozoa
Toxoplasma gondii

Spirochaetes
Treponema pallidum (syphilis)

Myositis
Bacteria
G-negatives
Fusobacterium
Pseudomonas aeruginosa
Staphylococcus spp.
Streptococcus spp.

Bone & joint infection
Acute arthritis
Native joint
Viruses
Arboviruses
EBV
HBV
Influenza
Mumps
Parvovirus
Rubella

Bacteria
Brucella
Escherichia coli
Neisseria gonorrhoeae
Neisseria meningitidis
Pseudomonas aeruginosa
Salmonella spp.
Staphylococcus aureus
Streptococcus pyogenes
Streptococcus group B β-haemolytic

Spriochaetes
Borrelia (Lyme disease)

Prosthetic joint
As above plus

Bacteria
Staphylococcus epidermidis

Fungi
Candida spp.

Reactive arthritis
Bacteria
Enteric pathogens

SIGNS & SYMPTOMS

Neisseria gonorrhoeae
Streptococcus pyogenes

Other organisms
Chlamydia trachomatis

Osteomyelitis

Bacteria
Escherichia coli
Haemophilus influenzae
Pseudomonas aeruginosa
Salmonella spp.
Staphylococcus aureus
Streptococcus pyogenes

Immunocompromised host

Acquired immune deficiency syndrome

Viruses
CMV
HSV
VZV

Mycobacteria
Mycobacterium avium
Mycobacterium intracellulare
Mycobacterium tuberculosis

Fungi
Candida spp.
Cryptococcus neoformans
Histoplasma spp.
Pneumocystis carinii

Protozoa
Cryptosporidium parvum
Microspora
Toxoplasma gondii

Alcohol abuse

Bacteria
Klebsiella
Streptococcus pneumoniae

Mycobacteria
Mycobacterium tuberculosis

Malnutrition

Bacteria
Salmonella spp.

Mycobacteria
Mycobacterium tuberculosis

Fungi
Pneumocystis carinii

Neonates

Bacteria
Escherichia coli
Listeria monocytogenes
Streptococcus β-haemolytic

Neutropaenia

Bacteria
Escherichia coli
Klebsiella
Pseudomonas aeruginosa
Staphylococcus epidermidis
Streptococcus viridans

Pregnancy

Viruses
Influenza
Polio
VZV

Bacteria
Streptococcus pneumoniae

Mycobacteria
Mycobacterium tuberculosis

Protozoa
Plasmodium spp.

Splenectomy

Bacteria
Babesia microti
Capnocytophaga canimorsus
Haemophilus influenzae
Neisseria meningitidis
Streptococcus pneumoniae
Streptococcus other haemolytic

PART 2: AETIOLOGICAL AGENTS

Viruses

DNA viruses	
Adenoviruses	Adenovirus
Hepadnaviruses	Hepatitis B Hepatitis D
Herpesviruses	Cytomegalovirus Epstein–Barr Herpes B Herpes simplex Human herpes virus 6 Human herpes virus 8 Varicella zoster
Papoviruses	Jamestown Canyon (JC) virus BK virus Papillomavirus
Parvoviruses	Parvovirus B19
Poxviruses	Molluscum contagiosum Smallpox
RNA viruses	
Arenaviruses	Lassa fever
Bunyaviruses	Hantavirus
Caliciviruses	Norwalk
Coronaviruses	Common cold Severe Acute Respiratory Syndrome (SARS)
Filoviruses	Ebola Marburg
Flaviviruses	Dengue Hepatitis C Japanese B encephalitis Tick-borne encephalitis Yellow fever
Orthomyxoviruses	Influenza
Paramyxoviruses	Measles Mumps Parainfluenza Respiratory syncitial virus

(*continued*)

Viruses *continued*

RNA viruses	
Picornaviruses	Coxsackie Echovirus Hepatitis A Polio Rhinovirus
Reoviruses	Rotavirus
Retroviruses	HIV 1 HIV 2 HTLV 1 HTLV 2
Rhabdoviruses	Rabies
Togaviruses	Alphaviruses – chikungunya, Sindbis, western equine encephalitis, Ross River, rubella
Unclassified	Hepatitis E

Bacteria	
Gram-positive cocci	Coagulase-negative staphylococci Group B streptococci Enterococci *Staphylococcus aureus* *Streptococcus pneumoniae* *Streptococcus pyogenes* *Streptococcus viridans*
Gram-positive bacilli	*Bacillus anthracis* *Bacillus cereus* *Corynebacterium diphtheriae* *Erysipelothrix* *Listeria monocytogenes*
Gram-negative cocci	*Moraxella* *Neisseria gonorrhoeae* *Neisseria meningitidis*
Gram-negative bacilli	*Acinetobacter* *Bartonella* spp. *Bordetella pertussis* *Brucella* spp. *Burkholderia* *Calymmatobacterium granulomatis* *Campylobacter jejuni* *Capnocytophaga*

	Enterobacteriaceae *Escherichia coli* *Francisella tularensis* *Gardnerella mobiluncus* *Gardnerella vaginalis* *Haemophilus influenzae* *Haemophilus* spp. *Helicobacter pylori* *Klebsiella* *Legionella pneumophila* *Pasteurella* spp. *Pseudomonas aeruginosa* *Salmonella* spp. *Shigella* spp. *Stenotrophomonas* *Streptobacillus moniliformis* *Vibrio cholerae* *Yersinia pestis* *Yersinia pseudotuberculosis*
Anaerobes	*Bacteroides* *Clostridium botulinum* *Clostridium perfringens* *Clostridium tetani* *Fusobacterium*

Mycobacteria
Mycobacterium avium
Mycobacterium leprae
Mycobacterium tuberculosis
Atypical mycobacteria

Fungi
Aspergillus
Blastomyces
Candida
Coccidioides
Cryptococcus
Dermatophytes

(*continued*)

Fungi *continued*

Histoplasma Mucormycosis	
Pneumocystis	
Sporothrix	

Protozoa

Babesia	
Cryptosporidium	
Entamoeba	*histolytica*
Giardia	
Leishmania	*donovani* *infantum* *mexicana*
Plasmodium	*falciparum* *malariae* *ovale* *vivax*
Toxoplasma	
Trichomonas	
Trypanosoma	brucei *cruzi*
Others	*isospora*

Helminths

Cestodes	tapeworms
Nematodes – intestinal	roundworms
Nematodes – tissue	*Dracunculus* filariasis *Onchocerca* trichinosis
Trematodes	*Schistosoma*
Visceral larva migrans	

Spirochaetes

Borrelia	burgdorferi *duttonii*
Leptospira	
Spirillum	*minimus*
Treponema	*pallidum* non-venereal

Other organisms

Chlamydia	*pneumoniae* *psittaci* *trachomatis*
Coxiella	*burnetii*
Ehrlichia	*chaffeensis* *phagocytophila*
Mycoplasma	*genitalium* *hominis* *pneumoniae*
Orientia	*tsutsugamushi*
Prions	
Rickettsia	*prowazekii* *rickettsii* *typhi*

Higher organisms

Actinomyces
Nocardia

Ectoparasites

Chiggers
Fleas
Flies

(continued)

AETIOLOGICAL AGENTS

PART 3: DISEASES

Notifiable diseases

Anthrax
Cholera
Diphtheria
Dysentery (amoebic or bacillary)
Encephalitis (acute)
Food poisoning
Haemorrhagic fevers (viral)
Hepatitis (viral)
Leprosy
Leptospirosis
Malaria
Measles
Meningitis
Meningococcal septicaemia (no meningitis)
Mumps
Ophthalmia neonatorum
Paratyphoid fever
Plague
Poliomyelitis (acute)
Rabies
Relapsing fever
Rubella
Scarlet fever
Smallpox
Tetanus
Tuberculosis
Typhoid fever
Typhus
Whooping cough
Yellow fever

D: caused by infection with *Actinomyces* spp.; characterised by indolent abscesses and chronic sinuses

A: *Actinomyces* spp. are G +ve anaerobes; part of normal buccal flora; often found in association with G −ves; disease most often due to *A. israelii*, *A. naeslundii*, *A. propionicum* & *A. viscosus*

A/R: recent dental work; poor dental hygiene; trauma; human bites; IUCD (rare)

E: worldwide; rare

H: constitutional upset; abscess or sinus formation; symptoms of local infiltration, e.g. haemoptysis

E: soft, relatively non-tender head & neck swellings → grow slowly → discharge externally; abscesses are cold; 25–50% involve an internal organ

P: abscess formation → cross fascial planes; may spread via blood

I: discharge/pus/sections – macroscopically for sulphur granules, H & E, silver or G stain for organisms, culture & sensitivity; blood cultures

M: prolonged high-dose antibiotics – penicillins, sulphonamides, erythromycin, chloramphenicol or tetracycline; surgical drainage & debridement

C: abdominal organ involvement (25–50%); myopericardial invasion (rare)

P: myopericardial invasion fatal, otherwise good; prevention – improved dental hygiene

Adenoviruses

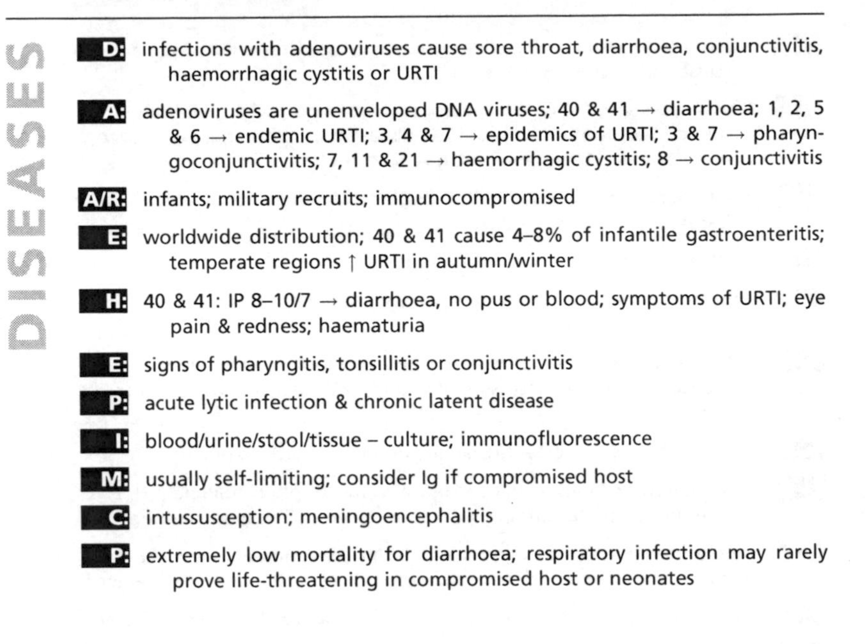

D: infections with adenoviruses cause sore throat, diarrhoea, conjunctivitis, haemorrhagic cystitis or URTI

A: adenoviruses are unenveloped DNA viruses; 40 & 41 → diarrhoea; 1, 2, 5 & 6 → endemic URTI; 3, 4 & 7 → epidemics of URTI; 3 & 7 → pharyngoconjunctivitis; 7, 11 & 21 → haemorrhagic cystitis; 8 → conjunctivitis

A/R: infants; military recruits; immunocompromised

E: worldwide distribution; 40 & 41 cause 4–8% of infantile gastroenteritis; temperate regions ↑ URTI in autumn/winter

H: 40 & 41: IP 8–10/7 → diarrhoea, no pus or blood; symptoms of URTI; eye pain & redness; haematuria

E: signs of pharyngitis, tonsillitis or conjunctivitis

P: acute lytic infection & chronic latent disease

I: blood/urine/stool/tissue – culture; immunofluorescence

M: usually self-limiting; consider Ig if compromised host

C: intussusception; meningoencephalitis

P: extremely low mortality for diarrhoea; respiratory infection may rarely prove life-threatening in compromised host or neonates

D: infections due to alphaviruses are named for the viruses that cause them – CHIK, Sindbis, W/E/VEE

A: alphaviruses are RNA viruses; spread by mosquito – CHIK & VEE *Aëdes* & *Culex* spp., Sindbis *Culex* spp., EEE *Culex* & *Culiseta* spp., WEE *Culex, Culiseta, Aëdes* & *Anopheles*

A/R: infants; young males; rural environment; malnutrition; occupational exposure

E: CHIK – Africa, India, S.E. Asia; Sindbis – Africa, India, tropical Asia, Australia; WEE – N.W. America; EEE – USA, C. & S. America; VEE – S. America

H: CHIK: IP 2–12/7 → biphasic illness Sindbis: (only occasionally overt disease in humans) → fever, rash, arthralgia, myalgia, malaise, headache W/E/VEE: IP 2–14/7 → short, sharp febrile attack – malaise, headache, stiffness, drowsiness → possible 2nd phase – excitability, somnolence, delirium, convulsions, paralysis, coma, EEE more severe than others

E: CHIK – rash is maculopapular, pruritic W/E/VEE – 2nd stage meningoencephalitic signs (stiff neck, drowsiness)

P: Ab neutralisation of virus after short illness; 2nd stage virus → nervous system → invades cells (grey matter) → destruction

I: virus can be isolated from blood in acute stage; Ab titres ↑ in convalescent sera

M: supportive

C: CHIK – arthralgia, arthritis
W/EEE – neurological complications in young children

P: majority recover completely → immunity; W/EEE 10% mortality, some permanent neurological sequelae in survivors; CHIK mortality up to 3% if $<$ 1-year-old or $>$ 50; prevention – avoid mosquito bites; vaccine available for selected populations

Amoebiasis

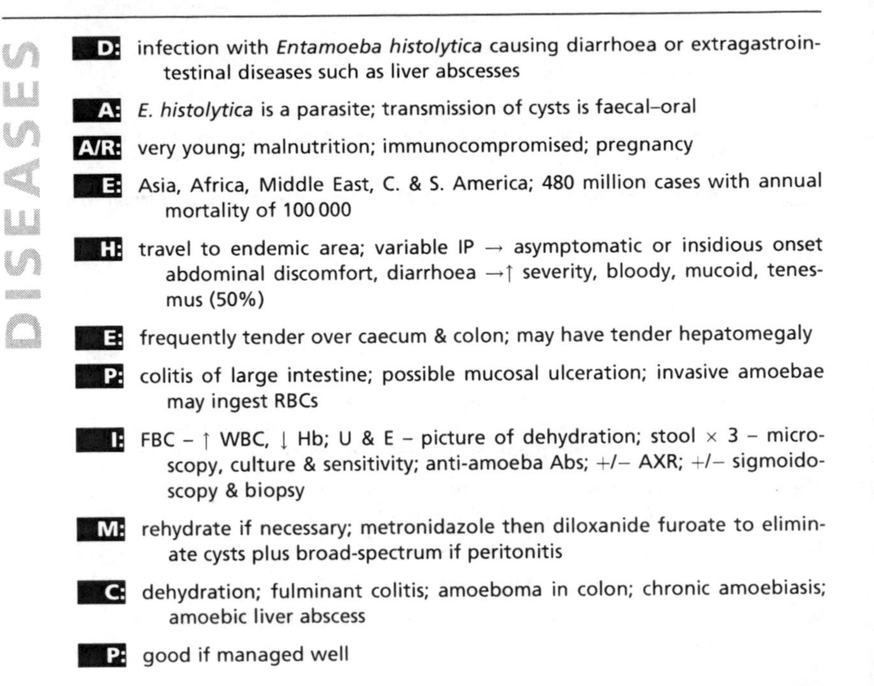

D: infection with *Entamoeba histolytica* causing diarrhoea or extragastrointestinal diseases such as liver abscesses

A: *E. histolytica* is a parasite; transmission of cysts is faecal–oral

A/R: very young; malnutrition; immunocompromised; pregnancy

E: Asia, Africa, Middle East, C. & S. America; 480 million cases with annual mortality of 100 000

H: travel to endemic area; variable IP $\rightarrow$ asymptomatic or insidious onset abdominal discomfort, diarrhoea $\rightarrow\uparrow$ severity, bloody, mucoid, tenesmus (50%)

E: frequently tender over caecum & colon; may have tender hepatomegaly

P: colitis of large intestine; possible mucosal ulceration; invasive amoebae may ingest RBCs

I: FBC – $\uparrow$ WBC, $\downarrow$ Hb; U & E – picture of dehydration; stool $\times$ 3 – microscopy, culture & sensitivity; anti-amoeba Abs; +/– AXR; +/– sigmoidoscopy & biopsy

M: rehydrate if necessary; metronidazole then diloxanide furoate to eliminate cysts plus broad-spectrum if peritonitis

C: dehydration; fulminant colitis; amoeboma in colon; chronic amoebiasis; amoebic liver abscess

P: good if managed well

D: anaerobes cause abscess formation as well as GI & RT disease

A: part of normal GIT & oral flora; *Fusobacterium necrophorum* causes Lemierre's syndrome, internal jugular vein septic thrombophlebitis; *Bacteroides fragilis, Clostridia, Peptostreptococcus* & *Prevotella* cause abscess formation, malabsorption, aspiration pneumonia & empyema

A/R: immunocompromised; starvation; alcoholism; diabetes; scleroderma; ileal bypass & blind loops of bowel; colonic cancer

E: worldwide

H: predisposing factors; abdominal or chest symptoms; sore throat (before Lemierre's)

E: signs of abscess; chest signs suggestive of pneumonia; tender neck; lymphadenopathy

P: abscess formation

I: blood/pus – microscopy, culture & sensitivity

M: surgical drainage; penicillin or metronidazole

C: necrotising jugular septic thrombophlebitis; septicaemia

P: mortality high in compromised & colon cancer

Anthrax

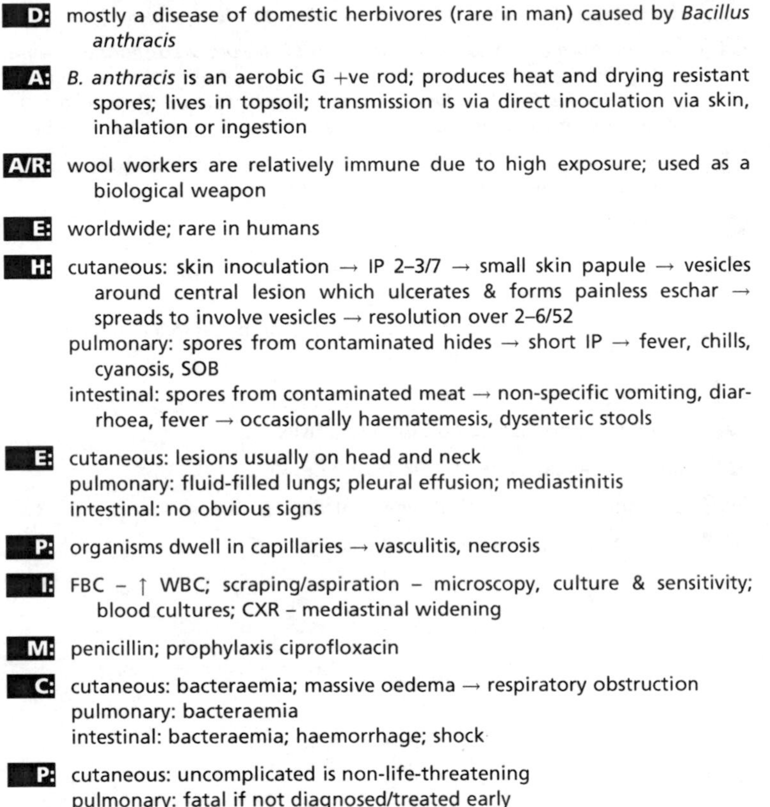

D: mostly a disease of domestic herbivores (rare in man) caused by *Bacillus anthracis*

A: *B. anthracis* is an aerobic G +ve rod; produces heat and drying resistant spores; lives in topsoil; transmission is via direct inoculation via skin, inhalation or ingestion

A/R: wool workers are relatively immune due to high exposure; used as a biological weapon

E: worldwide; rare in humans

H: cutaneous: skin inoculation → IP 2–3/7 → small skin papule → vesicles around central lesion which ulcerates & forms painless eschar → spreads to involve vesicles → resolution over 2–6/52
pulmonary: spores from contaminated hides → short IP → fever, chills, cyanosis, SOB
intestinal: spores from contaminated meat → non-specific vomiting, diarrhoea, fever → occasionally haematemesis, dysenteric stools

E: cutaneous: lesions usually on head and neck
pulmonary: fluid-filled lungs; pleural effusion; mediastinitis
intestinal: no obvious signs

P: organisms dwell in capillaries → vasculitis, necrosis

I: FBC – ↑ WBC; scraping/aspiration – microscopy, culture & sensitivity; blood cultures; CXR – mediastinal widening

M: penicillin; prophylaxis ciprofloxacin

C: cutaneous: bacteraemia; massive oedema → respiratory obstruction
pulmonary: bacteraemia
intestinal: bacteraemia; haemorrhage; shock

P: cutaneous: uncomplicated is non-life-threatening
pulmonary: fatal if not diagnosed/treated early
intestinal: most patients recover spontaneously
prevention: vaccine available

Aspergillosis

D: infection with *Aspergillus* spp. causing a spectrum of disease

A: *Aspergillus* spp. are fungi; important species are *A. fumigatus*, *A. flavus* & *A. niger*; spores found in soil, dust, decaying vegetable matter; infection is via inhalation of spores

A/R: immunocompromised (invasive disease); structural lung abnormality (aspergilloma); atopy (ABPA)

E: worldwide

H: ABPA: asthma, chronic cough
aspergilloma: cavitating lung disease in past, e.g. TB; intermittent cough; may develop haemoptysis
invasive: history of immunocompromise; symptoms of invasion

E: ABPA: wheeze

P: ABPA: hypersensitivity reaction
aspergilloma: formation of a fungal ball
invasive: invasion of lung, paranasal sinuses, CNS, kidney, bone, etc. by fungus

I: ABPA: CXR – more severe appearance than expected; peripheral shadowing aspergilloma: CXR/CT chest – SOL within a cavity with halo; sputum microscopy, culture & sensitivity
invasive: blood cultures; Ag detection; tissue biopsy

M: ABPA: steroids
aspergilloma: surgical excision
invasive: amphotericin or voriconazole; try and reverse/decrease immuno compromise

C: local invasion; bone erosion

P: high risk of fatality with invasive disease

Atypical mycobacteria

D: mostly incidental and opportunistic infections due to *Mycobacterium avium* & *Mycobacterium intracellulare* but also cutaneous granulomatous skin diseases

A: environmental saprophytes; Buruli ulcer – *Mycobacterium ulcerans*; swimming pool or fish tank granuloma – *Mycobacterium marinum*

A/R: predisposing lung lesion, e.g. COPD, old TB, CF; HIV; congenital immune deficiencies; ♂ > ♀

E: worldwide

H: pulmonary: insidious onset cough, weight loss in healthy/compromised
lymphadenopathy: < 5 years of age, healthy/compromised
post-inoculation: Buruli ulcer; swimming pool granuloma
disseminated: HIV or congenital immune deficiency

E: few signs

P: invasion of macrophages → immune response → granuloma formation

I: CXR; sputum/biopsy/excision – microscopy with Z–N stain, culture & sensitivity

M: antibiotics depend on site, severity, underlying condition, sensitivities, e.g. combinations of clarithromycin, doxycycline, rifampicin, ethambutol, isoniazid; surgical excision of lesion/lymph nodes/skin

C: dissemination

P: excellent in children with cervical adenitis; poor in immunocompromised

D: zoonotic infection with *Babesia* spp.

A: *Babesia* spp. are protozoan parasites of domestic & wild animals; transmission is via tick bite; mostly *B. bovis, B. microti, B. divergens*

A/R: splenectomy

E: rare; Europe mostly *B. divergens* spread by *Ixodes ricinus*; N. America mostly *B. microti* spread by *Ixodes dammini*

H: *divergens/bovis*: IP 1–4/52 → vague unwellness → fever, prostration, jaundice, fatigue
microti: IP 1–3/52 → mostly subclinical or anorexia, fatigue, fever, sweating, rigors, myalgia

E: *divergens/bovis*: splenectomy scar
microti: fever, mild splenomegaly +/– hepatomegaly

P: red cell infiltration & lysis

I: FBC – ↑ WBC, ↓ Hb; U & E ↑ urea; ↑ bilirubin (unconjugated); urinalysis – haematuria, proteinuria; blood film for parasites; consider IFA, PCR

M: *divergens*: anecdotal – diminazene (used in animals); co-trimoxazole + pentamidine; massive exchange transfusion + clindamycin + oral quinine
microti: quinine + clindamycin + blood or RBC exchange transfusion

C: ARF; haemolytic anaemia

P: *divergens/bovis*: untreated, splenectomised → death
microti: usually mild → spontaneous recovery

Bacillus cereus

D: cause of food poisoning with vomiting, diarrhoea or both

A: *Bacillus cereus* is a G +ve aerobe; can form spores; ubiquitous in soil; forms heat stable emetic toxin & heat labile enterotoxin

A/R: rice boiled in bulk and reheated, e.g. Chinese restaurants

E: worldwide; emetic toxin formed in food; enterotoxin formed in food but also in gut

H: emetic toxin: IP 1–5 h → vomiting; may have history of Chinese meal or similar
enterotoxin: IP 8–16 h → diarrhoea, abdominal pain

E: non-specific abdominal tenderness

P: non-specific

I: stool sample – microscopy, culture & sensitivity; also test food samples

M: supportive

C: dehydration

P: symptoms generally do not persist beyond 24 h

Bacterial vaginosis

D: syndrome characterised by vaginal discharge & disruption of normal vaginal flora (↑ anaerobes, ↓ lactobacilli)

A: increase in anaerobes; mainly *Gardnerella vaginalis* & *Mycoplasma hominis*

A/R: more prevalent among ♀ with multiple partners (but not a proven STI)

E: worldwide

H: white discharge +/– odour

E: white homogeneous discharge (90%); unpleasant odour (90%)

P: shift in the balance of vaginal flora; no inflammation

I: discharge – G stain; slide preparation with 10% potassium hydroxide (fishy smell); wet preparation microscopy for clue cells

M: metronidazole

C: premature labour; chorioamnionitis; postpartum endometritis; ? PID

P: recurrence common

Bartonellosis

D: infection with *Bartonella bacilliformis*; also known as Oroya fever, Guaitara fever, Carrión's disease, Verruga peruana

A: *B. bacilliformis* is a G −ve bacillus; pleomorphic; transmission is via sandfly bites

A/R: splenectomy

E: outbreaks only occur between 9 & 16°S @ 800–3000 m altitude – Colombia, Peru, Ecuador; *Bartonella*-like infection in Thailand, Niger, Sudan, E. USA, Pakistan

H: commonly asymptomatic carriage
Oroya fever: IP 3/52 → insidious start → irregular remitting fever, severe bone pain, fatigue (due to anaemia)
Verruga peruana: sequel to Oroya (30–40/7 later)

E: granulomatous skin eruptions; splenomegaly (if not splenectomised)

P: invasion of RBCs → multiplication, destruction of RBCs; also invasion of reticuloendothelial cells → lymph gland hyperplasia, necrotic foci in liver, spleen, bone marrow; also parasitise endothelium

I: FBC – ↑ WBC, ↓ Hb, ↔ MCV; blood film; Verruga smear; blood culture & sensitivity

M: chloramphenicol or penicillin, tetracycline, co-trimoxazole

C: salmonellosis; thromboses; pleurisy; parotitis; meningoencephalitis; CNS involvement

P: Oroya has 10–40% mortality; CNS involvement has high mortality; recovery from any form gives lasting immunity

Blastomycosis

D: local or systemic infection with *Blastomyces dermatitidis*

A: *B. dermatitidis* is a dimorphic fungus; transmission via inhalation of yeast phase

A/R: disseminated disease in immunocompromised

E: mainly USA & Canada, but also Africa, India, Middle East

H: may be asymptomatic
skin lesions: initial single nodule → crusted plaques, ulcers, abscesses
chronic pulmonary: cough
disseminated: cough; skin lesions on face and forearms

E: skin signs; chest signs

P: non-caseating granuloma formation

I: sputum/scrapings – microscopy & culture (mould @ room temperature, yeast @ 37°C hence diamorphine)

M: itraconazole or ketoconazole or amphotericin for life-threatening illness

C: lytic bone lesions (especially axial skeleton); GU tract disease (especially epididymitis)

P: curative in immunocompetent

Botulism

D: infection with *Clostridium botulinum* causing a paralytic illness

A: *C. botulinum* is an anaerobic bacterium; can form heat-resistant spores that germinate to produce a neurotoxin; widespread in soil; transmission via contaminated food

A/R: consumption of preserved foods which have been inadequately heat-treated

E: rare in UK (last outbreak 1989)

H: IP 12–36 h → vomiting, fatigue, visual disturbance (ocular muscle paralysis), swallowing/speech disturbance (bulbar muscle paralysis) → flaccid limb/trunk paralysis

E: flaccid paralysis; sensation intact

P: neurotoxin mediated

I: blood/stool samples – presence of toxin

M: supportive; antitoxin

C: respiratory failure

P: 50% death from respiratory failure; 50% gradual but complete recovery

D: systemic infection with *Brucella* spp.

A: *Brucella* spp. are G –ve facultative intracellular bacteria; main species in man are *B. abortus* (from cattle), *B. melitensis* (goats), *B. suis* (pigs), *B. canis* (dogs); transmission via contaminated animal products

A/R: unpasteurised milk or cheese; infected meat; occupational exposure (farmworkers, laboratory staff); HIV/AIDS

E: 500 000 cases worldwide p.a.; 20–30 in UK p.a.

H: IP 1–28/7 → sweats, high fever (undulant), rigors, myalgia, malaise, arthritis/arthralgia (monoarticular, large joint); +/– low back pain, sciatica; headache, irritability, insomnia, confusion

E: lymphadenopathy; hepatosplenomegaly

P: spread to bloodstream via lymphatics → bacteraemia → multiplication & localisation → granulomatous response → necrosis & abscess formation

I: FBC – ↔ WBC, ↓ Hb, ↔ MCV; LFT – mildly deranged; ↑ ESR; blood/bone marrow culture; Abs – ELISA or agglutination

M: streptomycin + doxycycline + gentamicin if hospitalised; rifampicin + co-trimoxazole if < 8 years old or pregnant

C: arthritis, sacroiliitis, vertebral osteomyelitis (10–30%), toxic course; meningitis, encephalitis, peripheral neuritis; endocarditis, myocarditis, pericarditis; granulomatous hepatitis; epididymo-orchitis; bronchopneumonia, pleurisy

P: most infections are mild and self-limiting over 2–3/52; prevention by animal vaccination, pasteurisation

Campylobacter jejuni

D: food-borne cause of diarrhoeal disease

A: *Campylobacter* spp. are G –ve bacilli; disease mostly due to *Campylobacter jejuni* (but may also be *C. fetus* in immunocompromised); transmission via contaminated food (especially poultry), milk, water & pets

A/R: children & young adults

E: most common bacterial diarrhoea in UK

H: IP 2/7 → fever, myalgia, abdominal pain → diarrhoea (large volume, watery, offensive) → small volume +/– blood +/– mucus for 1/52

E: tender abdomen

P: invasion of intestinal mucosa (especially ileum and colon) → inflammatory response

I: stool – microscopy, culture & sensitivity

M: rehydration; erythromycin or ciprofloxacin in severe cases

C: ileitis, colitis; bacteraemia; GBS; cholecystitis; erythema nodosum; reactive arthritis

P: normally resolves spontaneously in 1/52

D: infection with *Candida* spp.

A: *Candida* spp. are yeasts; *C. albicans* is a normal commensal of vagina, mouth & GIT but causes vulvovaginal & oropharyngeal disease and can also cause 2° infection of nappy rash; also *C. dubliniensis, C. glabrata, C. guilliermondii, C. krusei, C. lusitaniae, C. parapsilosis* & *C. pseudotropicalis* increasingly common in immunocompromised hosts

A/R: vulvovaginal: pregnancy; OCP; antimicrobial therapy; immunocompromised
oropharyngeal: extremes of age; immunocompromised; steroids; diabetes

E: worldwide

H: vulvovaginal: pruritis vulvae; whitish discharge
oropharyngeal: oral discomfort
systemic: neutropaenia; major surgery; long-term IV feeding; HIV/AIDS

E: white plaques; erythema +/– oedema

P: fungal plaque formation; inflammation

I: vulvovaginal: swab – wet slide preparation
all: scrapings/swabs direct microscopy & culture

M: vulvovaginal: topical nystatin cream or oral fluconazole
oropharyngeal: spray/mouthwash/pastilles/tablets of amphotericin B, nystatin or ketoconazole may need oral azole in compromised hosts
systemic: IV amphotericin or azole

C: dissemination

P: local infection normally resolves with treatment; systemic infection may prove fatal

Capnocytophaga

D: infections with *Capnocytophaga* spp. cause suppurative infections

A: *Capnocytophaga* spp. are G −ve bacilli; commensals of oral cavity; transmission usually via human or animal bite; most commonly *C. canimorsus* or *C. cynodegmi*

A/R: splenectomy; immunocompromised; alcohol abuse; steroids; working with animals

E: worldwide

H: history of bite; swelling, inflammation at site of bite

E: signs of suppurative infection

P: suppuration

I: pus/blood – G stain, culture & sensitivity

M: clean wound; co-amoxiclav

C: septicaemia; arthritis; keratitis; mucositis

P: higher mortality in compromised host

D: chickenpox is a 1° infection with varicella; shingles is a reactivation of the virus with 2° zoster infection

A: varicella zoster belongs to the herpes virus family

A/R: ↑ mortality in Hodgkin's & non-Hodgkin's lymphoma, AIDS, leukaemia, immunosuppression, pregnancy, neonates, elderly

E: endemic worldwide; affects 90% of children < 10 years old

H: chickenpox: IP 14–15/7 → malaise, mild fever, rash (starting on head)
shingles: dermatomal pain preceding lesion formation

E: chickenpox: characteristic rash with crusty vesicles; spreads from head
shingles: characteristic unilateral rash with crusting & eruption

P: mononuclear infiltration; inflammation

I: serology – ELISA, CF, IFA, radioimmunoassays

M: antipyretics & pain relief (not aspirin); consider acyclovir +/– Ig depending on host status

C: chickenpox: 2° bacterial infection; pneumonitis; CNS complications especially cerebellar ataxia
shingles: encephalitis; post-herpetic neuralgia; ophthalmic zoster; autonomic zoster, motor zoster

P: good in competent host; prevention – vaccine now licensed in UK & N. America

Chlamydiae

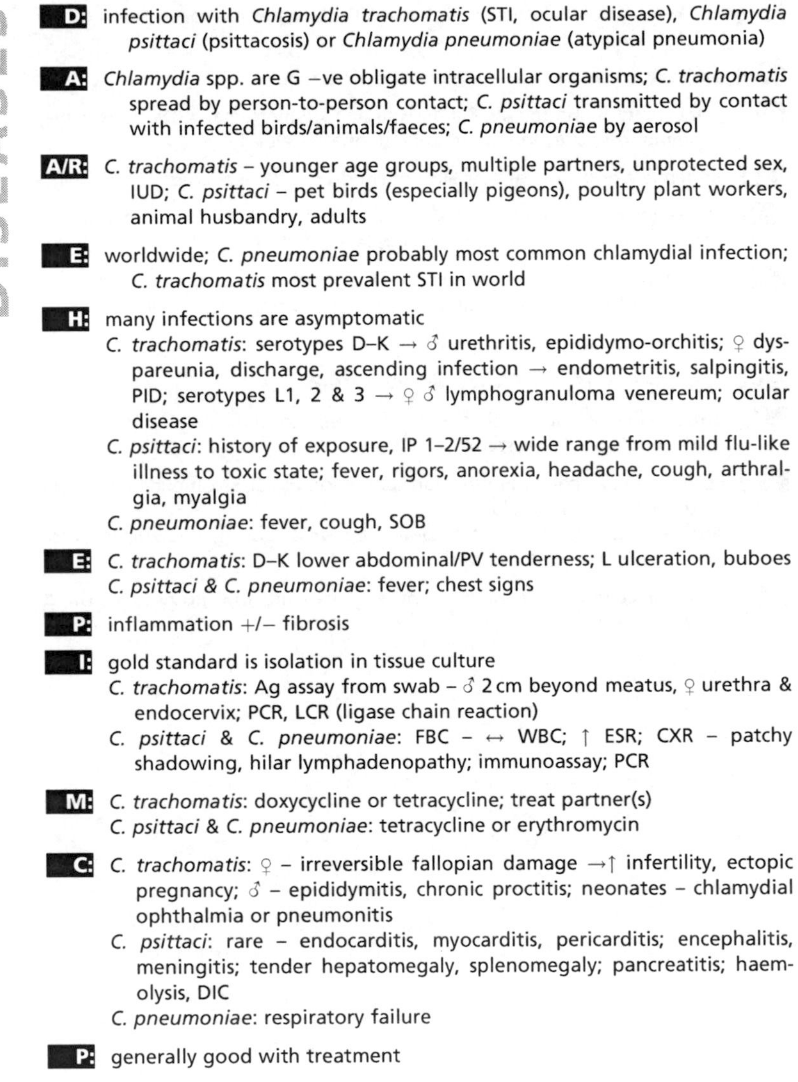

D: infection with *Chlamydia trachomatis* (STI, ocular disease), *Chlamydia psittaci* (psittacosis) or *Chlamydia pneumoniae* (atypical pneumonia)

A: *Chlamydia* spp. are G –ve obligate intracellular organisms; *C. trachomatis* spread by person-to-person contact; *C. psittaci* transmitted by contact with infected birds/animals/faeces; *C. pneumoniae* by aerosol

A/R: *C. trachomatis* – younger age groups, multiple partners, unprotected sex, IUD; *C. psittaci* – pet birds (especially pigeons), poultry plant workers, animal husbandry, adults

E: worldwide; *C. pneumoniae* probably most common chlamydial infection; *C. trachomatis* most prevalent STI in world

H: many infections are asymptomatic

C. trachomatis: serotypes D–K → ♂ urethritis, epididymo-orchitis; ♀ dyspareunia, discharge, ascending infection → endometritis, salpingitis, PID; serotypes L1, 2 & 3 → ♀ ♂ lymphogranuloma venereum; ocular disease

C. psittaci: history of exposure, IP 1–2/52 → wide range from mild flu-like illness to toxic state; fever, rigors, anorexia, headache, cough, arthralgia, myalgia

C. pneumoniae: fever, cough, SOB

E: *C. trachomatis*: D–K lower abdominal/PV tenderness; L ulceration, buboes

C. psittaci & C. pneumoniae: fever; chest signs

P: inflammation +/– fibrosis

I: gold standard is isolation in tissue culture

C. trachomatis: Ag assay from swab – ♂ 2 cm beyond meatus, ♀ urethra & endocervix; PCR, LCR (ligase chain reaction)

C. psittaci & *C. pneumoniae*: FBC – ↔ WBC; ↑ ESR; CXR – patchy shadowing, hilar lymphadenopathy; immunoassay; PCR

M: *C. trachomatis*: doxycycline or tetracycline; treat partner(s)

C. psittaci & *C. pneumoniae*: tetracycline or erythromycin

C: *C. trachomatis*: ♀ – irreversible fallopian damage →↑ infertility, ectopic pregnancy; ♂ – epididymitis, chronic proctitis; neonates – chlamydial ophthalmia or pneumonitis

C. psittaci: rare – endocarditis, myocarditis, pericarditis; encephalitis, meningitis; tender hepatomegaly, splenomegaly; pancreatitis; haemolysis, DIC

C. pneumoniae: respiratory failure

P: generally good with treatment

Cholera

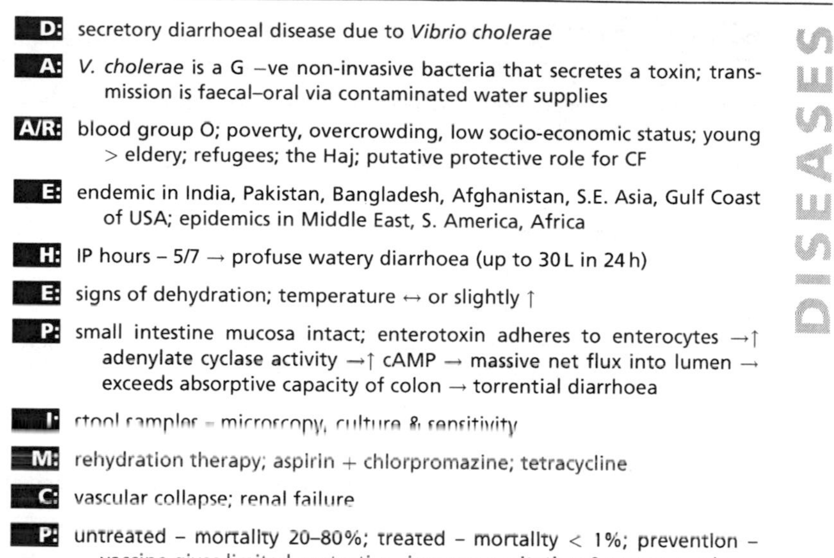

D: secretory diarrhoeal disease due to *Vibrio cholerae*

A: *V. cholerae* is a G −ve non-invasive bacteria that secretes a toxin; transmission is faecal–oral via contaminated water supplies

A/R: blood group O; poverty, overcrowding, low socio-economic status; young > eldery; refugees; the Haj; putative protective role for CF

E: endemic in India, Pakistan, Bangladesh, Afghanistan, S.E. Asia, Gulf Coast of USA; epidemics in Middle East, S. America, Africa

H: IP hours – 5/7 → profuse watery diarrhoea (up to 30 L in 24 h)

E: signs of dehydration; temperature ↔ or slightly ↑

P: small intestine mucosa intact; enterotoxin adheres to enterocytes →↑ adenylate cyclase activity →↑ cAMP → massive net flux into lumen → exceeds absorptive capacity of colon → torrential diarrhoea

I: stool samples – microscopy, culture & sensitivity

M: rehydration therapy; aspirin + chlorpromazine; tetracycline

C: vascular collapse; renal failure

P: untreated – mortality 20–80%; treated – mortality < 1%; prevention – vaccine gives limited protection, improve sanitation & water supply

Coccidiomycosis

D: infection with *Coccidioides immitis* causing pneumonitis

A: *C. immitis* is a white mould yeast; lives in soil; spread by aerosol

A/R: progression more likely in American Indians, black people, mestizos; HIV/AIDS; pregnancy

E: USA, C. America, Colombia, Venezuela, Argentina, Paraguay

H: up to 70% asymptomatic or fever, weight loss, cough, chest pain, arthralgia, conjunctivitis

E: fever, erythema nodosum/multiforme, chest signs

P: granulomatous response to fungi

I: smears/biopsies/sputum – microscopy & culture; serology; CXR – focal consolidation/pleural effusion/hilar lymphadenopathy

M: ketoconazole, itraconazole or fluconazole PO, or amphotericin B IV

C: dissemination to joints, meninges, skin, other organs

P: response to widely disseminated disease is poor; otherwise prognosis good

Common cold

D: nasal or nasopharyngeal stinging, blockage, discharge

A: mostly due to coronavirus & rhinovirus; transmission by aerosol & droplet

A/R: ubiquitous

E: more common in winter

H: general malaise; runny nose; stuffiness

E: mild pharyngitis

P: coronavirus: destruction of ciliated epithelia; rhinovirus: little mucosal damage, mostly due to release of local inflammatory mediators

I: clinical diagnosis

M: supportive; treat 2° infections only

C: 2° bacterial infections. Severe Acute Respiratory Syndrome (SARS) is an epidemic viral respiratory illness caused by a novel coronavirus with a mortality of around 8%.

P: excellent; suffer 1–6 p.a. (poor immunity)

Coxsackie & Echoviruses

D: coxsackie & echoviruses cause non-specific febrile illnesses

A: both are enteroviruses; both cause non-specific illnesses, aseptic meningitis & encephalitis; coxsackie may cause HUS; group A coxsackie also causes hand–foot–mouth disease; group B coxsackie also causes pericarditis & Bornholm's disease

A/R: infants

E: worldwide

H: non-specific febrile illness

E: fever; pleural or pericardial rub; signs of meningitis

P:

I: CSF/stool sample – culture & PCR

M: supportive; ? role for pleconaril

C: febrile convulsions; paralysis

P: recovery normally complete

D: systemic infection caused by *Cryptococcus neoformans*

A: *C. neoformans* is an encapsulated yeast; *var. neoformans* more likely to cause disease in compromised hosts, *var. gattii* in competent hosts; dwells in soil & pigeon excreta

A/R: immunocompromised

E: Africa, Far East, Papua New Guinea, Australia, USA – incidence varies

H: pulmonary: cough, chest pain, fever
CNS: neck stiffness, headache

E: pulmonary: chest signs CNS: confusion, drowsiness, photophobia, cranial nerve palsies

P:

I: smears/CSF/sputum – microscopy with Indian ink stain & culture; Ag detection

M: competent: amphotericin B + flucytosine
compromised: less clear – as above + long-term fluconazole, repeat lumbar puncture to control CSF pressure

C: dissemination to liver, spleen, kidney, bone; chronic meningitis

P: mortality low, even in compromised hosts

D: cause diarrhoea in competent hosts & act as opportunistic infections in compromised hosts

A: protozoa; transmission is via faecal–oral route

A/R: childhood; day care facilities; HIV/AIDS

E: endemic worldwide

H: may be asymptomatic
diarrhoea: self-limiting in competent; chronic in compromised

E: dehydration

P:

I: stool – microscopy with auramine, modified Z–N or mAbs for oocysts; serology – ELISA; biopsy – EM

M: fluid replacement; co-trimoxazole for *Isospora* & *Cyclospora*

C: chronic diarrhoea

P: 50% relapse rate in *Isospora*

D: dengue fever, dengue shock syndrome & dengue haemorrhagic fever are caused by dengue virus

A: most important human arbovirus; 4 antigenically distinct serotypes DEN1, DEN2, DEN3 & DEN4; transmitted person to person by *Aëdes* (day-biting) mosquitoes; cross-reactive Ags with JBE & WNV

A/R: DHF – ? existing infection with a different serotype

E: between 30°N & 40°S; endemic in S.E. Asia (1–4), Pacific (1–3), E. & W. Africa (1–4, no DHF), Caribbean (1–4), Americas (1–4)

H: DF: may be asymptomatic or IP 5–8/7 → undifferentiated fever – infants/children → fever; older children/adults → fever, myalgia, arthralgia; also anorexia, sore throat, eye pain, abdominal pain, skin pain/itching
DHF: abrupt onset fever, headache, flushing, anorexia, vomiting, abdominal pain, petechiae

E: DF: high temperature; weakness, prostration; rash +/− dermal hyperaesthesia; lymphadenopathy
DHF: temperature 40–41°C; febrile convulsions; hepatomegaly; lymphadenopathy; +ve tourniquet test & bruising

P: virus → lymph nodes → reticuloendothelial system → multiplication → blood DHF also plasma leakage (may be immune-mediated)

I: FBC – ↓ platelets; serology – ELISA or MACELISA; Abs ↑× 4 from admission to 3–5/7 later, 3rd specimen @ 2–3/52

M: DF: symptomatic supportive
DHF: symptomatic supportive; replace plasma; antipyretics (not aspirin)

C: DSS; haemorrhagic complications, e.g. epistaxis, GI bleeding; encephalitic signs; Reye's syndrome

P: 2% mortality for DHF & DSS

Dermatophytes

D: cause superficial fungal infections – tinea corporis (trunk), capitis (scalp), cruris (groin), pedis (feet), imbricata & onychomycosis (nails)

A: mould fungi; mainly *Trichophyton* spp., *Microsporum* spp. & *Epidermophyton* spp.

A/R: climate; humidity of skin surface

E: common worldwide

H: itching & scaling

E: round scaly plaques with pronounced edge containing scales & papules

P: fungi adhere to stratum corneum → attack keratin

I: skin scrapings/hair/nail samples – microscopy, culture & sensitivity

M: corporis – topical clotrimazole or econazole, oral griseofulvin or itraconazole if extensive; capitis – oral griseofulvin; pedis – topical clotrimazole or other; onycholysis – oral terbinafine or itraconazole

C: hair loss if scarring

P: relapse rates high for onycholysis

Diphtheria

D: acute infection of tonsils, pharynx, larynx or nose caused by *Corynebacterium diphtheriae* with severe complications due to toxin production

A: *C. diphtheriae* is a G +ve pleomorphic bacterium; transmission is by droplets or direct contact with secretions

A/R: childhood; unvaccinated

E: significant problem in many developing countries; developed countries have much lower incidence due to vaccination

H: fever, nasal discharge, hoarseness, cough, tonsillitis

E: cervical lymphadenopathy; grey-white pseudomembrane @ various possible locations

P: local destruction of epithelial cells & distant effects on heart, kidneys, peripheral nerves by exotoxin

I: swabs/blood cultures/discharge – microscopy & culture, test for toxin production

M: emergency tracheostomy; ECG monitoring; antitoxin; penicillin G or erythromycin

C: 65% have some cardiac involvement, e.g. arrhythmia; 10% myocarditis; 7–10% neuritis, paralysis

P: myocarditis has 50% mortality; prevention – vaccination, contact tracing & prophylaxis

Ectoparasites

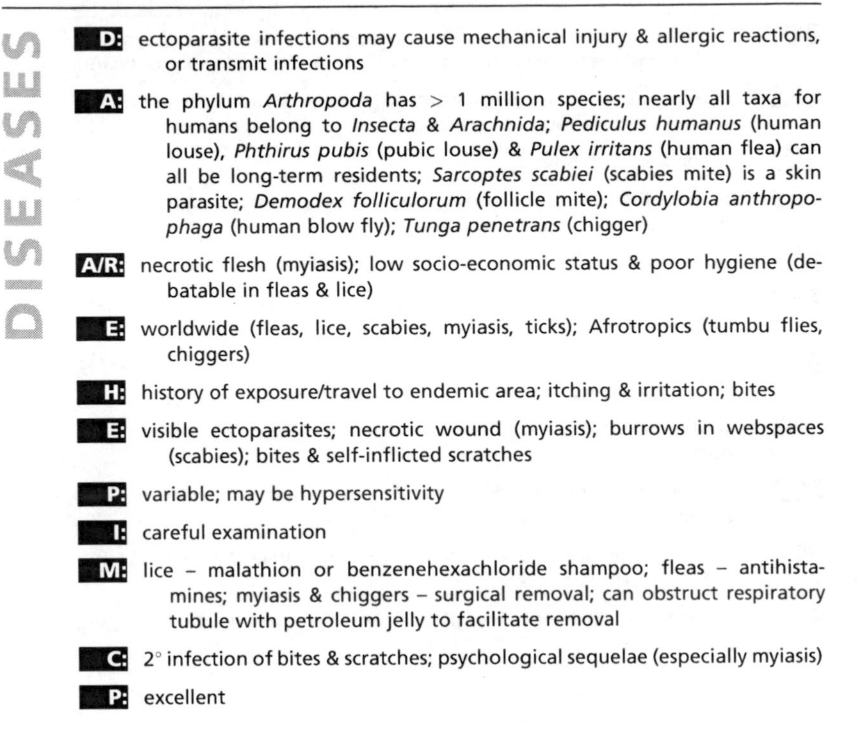

D: ectoparasite infections may cause mechanical injury & allergic reactions, or transmit infections

A: the phylum *Arthropoda* has > 1 million species; nearly all taxa for humans belong to *Insecta* & *Arachnida*; *Pediculus humanus* (human louse), *Phthirus pubis* (pubic louse) & *Pulex irritans* (human flea) can all be long-term residents; *Sarcoptes scabiei* (scabies mite) is a skin parasite; *Demodex folliculorum* (follicle mite); *Cordylobia anthropophaga* (human blow fly); *Tunga penetrans* (chigger)

A/R: necrotic flesh (myiasis); low socio-economic status & poor hygiene (debatable in fleas & lice)

E: worldwide (fleas, lice, scabies, myiasis, ticks); Afrotropics (tumbu flies, chiggers)

H: history of exposure/travel to endemic area; itching & irritation; bites

E: visible ectoparasites; necrotic wound (myiasis); burrows in webspaces (scabies); bites & self-inflicted scratches

P: variable; may be hypersensitivity

I: careful examination

M: lice – malathion or benzenehexachloride shampoo; fleas – antihistamines; myiasis & chiggers – surgical removal; can obstruct respiratory tubule with petroleum jelly to facilitate removal

C: 2° infection of bites & scratches; psychological sequelae (especially myiasis)

P: excellent

Ehrlichosis

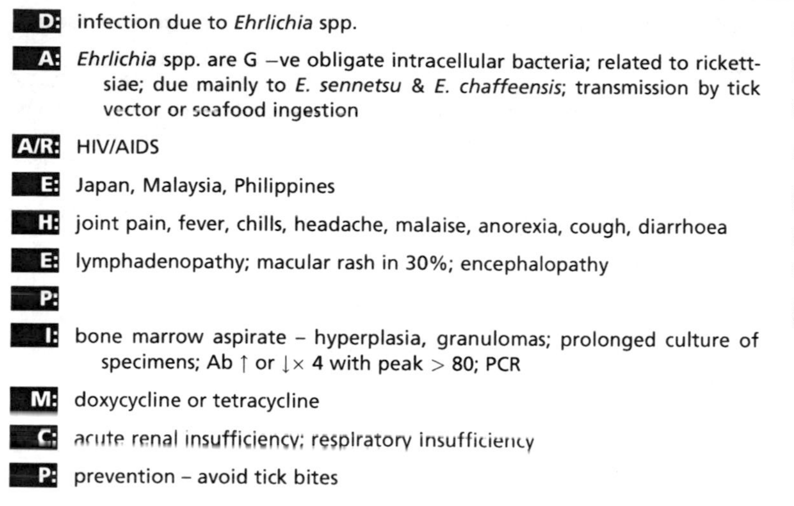

D: infection due to *Ehrlichia* spp.

A: *Ehrlichia* spp. are G −ve obligate intracellular bacteria; related to rickettsiae; due mainly to *E. sennetsu* & *E. chaffeensis*; transmission by tick vector or seafood ingestion

A/R: HIV/AIDS

E: Japan, Malaysia, Philippines

H: joint pain, fever, chills, headache, malaise, anorexia, cough, diarrhoea

E: lymphadenopathy; macular rash in 30%; encephalopathy

P:

I: bone marrow aspirate – hyperplasia, granulomas; prolonged culture of specimens; Ab ↑ or ↓× 4 with peak > 80; PCR

M: doxycycline or tetracycline

C: acute renal insufficiency; respiratory insufficiency

P: prevention – avoid tick bites

Filariasis, Dracunculiasis, Trichinosis

D: infection with tissue-dwelling nematodes causing a spectrum of diseases

A: **filariasis** – *Wuchereria bancrofti, Brugia malayi, Brugia timori* are transmitted by *Aëdes, Anopheles, Culex* & *Mansoni* & cause filarial fever, adenolymphangitis & lymphatic filariasis; *Onchocerca volvulus* is transmitted by *Simulium* & causes onchocerciasis & river blindness; *Loa loa* is transmitted by *Chrysops* & causes Calabar swelling; *Mansonella streptocerca* causes minor skin disease
dracunculiasis – *Dracunculus medinensis* (guinea worm) is transmitted by ingestion of water contaminated by larvae
trichinosis – *Trichinella spiralis* larvae are transmitted via infected meat

A/R: **filariasis** – multiple coinfections; fast-flowing water (onchocerciasis)
dracunculiasis – poverty; poor water supply
trichinosis – eating poorly prepared pork, boar or bear

E: **filariasis** – tropics; $>$ 100 million infected
dracunculiasis – Africa & Asia; 1 million cases p.a.
trichinosis – C. & E. Europe, C., S. & N. America, Africa, Asia & Arctic

H: **filariasis** – history of exposure; may be asymptomatic; may have skin lesions, ocular lesions or oedema
dracunculiasis – history of exposure; IP 9–14/12 → swelling & pain @ site of eruption of migrating female
trichinosis – history of ingestion of meat; may be asymptomatic; IP 1/52 → abdominal pain, nausea, vomiting, diarrhoea, fever, sweating; 1/52 → oedema, fever, myalgia, rash; 1/52 → recovery

E: **filariasis** – skin lesions (similar to leprosy); characteristic ocular lesions (river blindness); lymphoedema
dracunculiasis – worm just under dermis of trunk/limbs; blister → ulcer → abscess; worm protruding from lesion
trichinosis – rash is fine, macular

P: **filariasis** – partially immune-mediated
dracunculiasis – delayed hypersensitivity
trichinosis – hypersensitivity

I: **filariasis** – FBC – ↑ eosinophils; biopsy for parasite identification; immunological tests unreliable
dracunculiasis – FBC ↑ eosinophils; saline preparation of exudate for larvae; ELISA; skin test
trichinosis – FBC – ↑ eosinophils; LFT – mildly deranged AST & ALT; skin biopsy for Ag; slide flocculation test; muscle biopsy for larvae

M: **filariasis** – ivermectin or diethylcarbamazine citrate
dracunculiasis – surgical/manual extraction of worms; analgesia; anti-inflammatories; treat 2° infections
trichinosis – bedrest & salicylates; prednisolone if myocarditis or severe myalgia; consider mebendazole

C: **filariasis** – blindness (onchocerciasis); hydrocele, lymphoedema, elephantiasis, tropical eosinophilia (lymphatic filariasis)
dracunculiasis – 2° bacterial infections causing cellulitis, abscesses, gangrene; repeated infection may result in joint ankylosis
trichinosis – myocardial, lung or CNS involvement

P: **filariasis** – generally good if treated effectively; prevention – avoid larval sources, eradication & education campaigns
dracunculiasis – 1% mortality; prevention – boil/filter water, develop safe water supply, education
trichinosis – good; prevention – adequate freezing/cooking of meat; improved animal rearing

Gangrene

D: rapidly developing & spreading infection of muscle and soft tissues due to toxin-producing *Clostridium* spp.

A: *Clostridium* spp. are G +ve spore forming anaerobic bacteria; mostly due to *C. perfringens*; occurs naturally in soil & GIT; transmission is via direct inoculation

A/R: proximity to faecal sources of bacteria, e.g. hip surgery; wound contamination, e.g. shrapnel, dirt; tissue necrosis

E: worldwide

H: IP < 4/7, often < 1/7, sometimes < 6 h → pain @ wound site →↑ pain, fever

E: swelling, skin discolouration, thin serous ooze → haemorrhagic vesicles, necrosis, crepitus, ↑ pulse, ↑ wound pain

P:

I: diagnose & treat on clinical grounds as emergency; discharge – microscopy & culture (although 30% of wounds will be colonised anyway); XR to look for gas

M: surgical debridement; penicillin; hyperbaric O_2

C: amputation; disfigurement

P: fatal if untreated; may need repeated debridement +/– amputation; prevention – antitoxin, perioperative penicillin or metronidazole

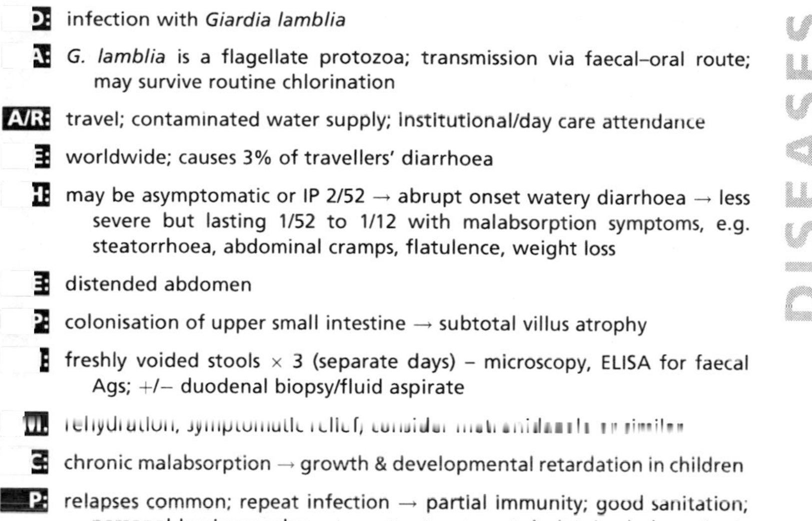

D: infection with *Giardia lamblia*

A: *G. lamblia* is a flagellate protozoa; transmission via faecal–oral route; may survive routine chlorination

A/R: travel; contaminated water supply; institutional/day care attendance

E: worldwide; causes 3% of travellers' diarrhoea

H: may be asymptomatic or IP 2/52 → abrupt onset watery diarrhoea → less severe but lasting 1/52 to 1/12 with malabsorption symptoms, e.g. steatorrhoea, abdominal cramps, flatulence, weight loss

E: distended abdomen

P: colonisation of upper small intestine → subtotal villus atrophy

I: freshly voided stools × 3 (separate days) – microscopy, ELISA for faecal Ags; +/– duodenal biopsy/fluid aspirate

M: rehydration, symptomatic relief; consider metronidazole or similar

C: chronic malabsorption → growth & developmental retardation in children

P: relapses common; repeat infection → partial immunity; good sanitation; personal hygiene; adequate water treatment; isolate/exclude patients if in hospital/work or school

Glandular fever

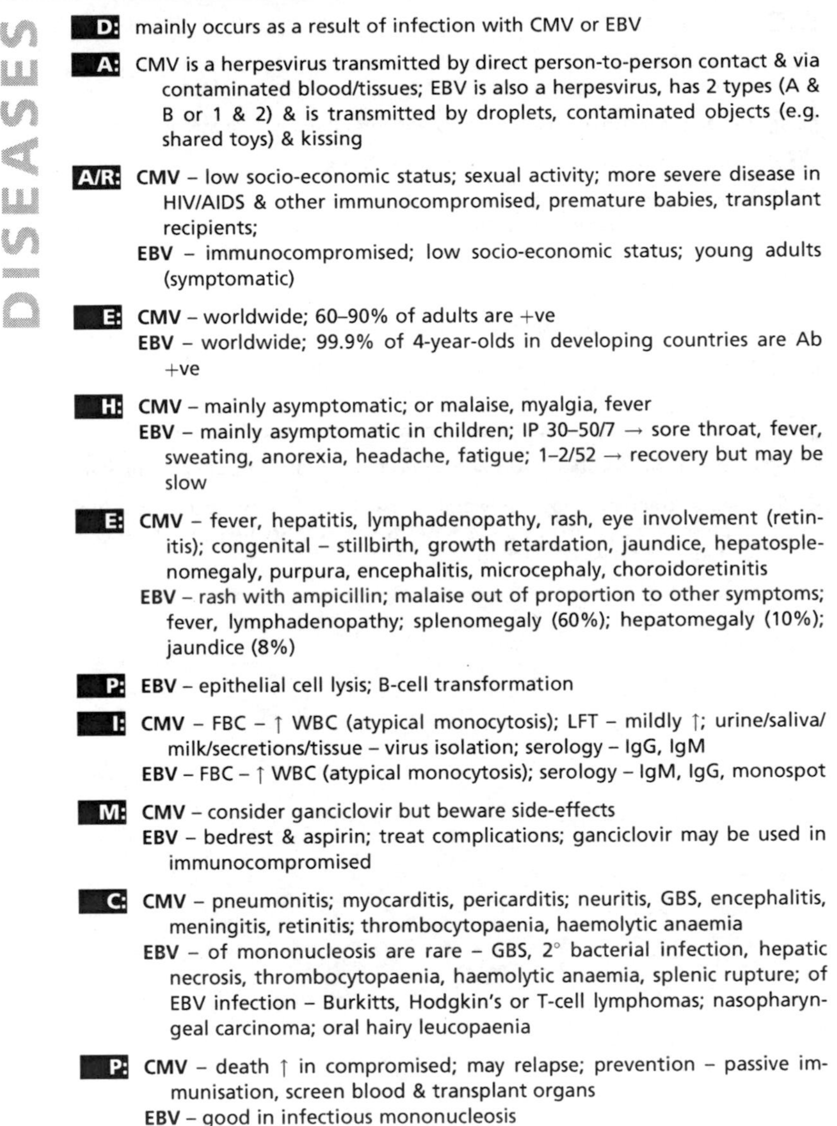

D: mainly occurs as a result of infection with CMV or EBV

A: CMV is a herpesvirus transmitted by direct person-to-person contact & via contaminated blood/tissues; EBV is also a herpesvirus, has 2 types (A & B or 1 & 2) & is transmitted by droplets, contaminated objects (e.g. shared toys) & kissing

A/R: **CMV** – low socio-economic status; sexual activity; more severe disease in HIV/AIDS & other immunocompromised, premature babies, transplant recipients;
EBV – immunocompromised; low socio-economic status; young adults (symptomatic)

E: **CMV** – worldwide; 60–90% of adults are +ve
EBV – worldwide; 99.9% of 4-year-olds in developing countries are Ab +ve

H: **CMV** – mainly asymptomatic; or malaise, myalgia, fever
EBV – mainly asymptomatic in children; IP 30–50/7 → sore throat, fever, sweating, anorexia, headache, fatigue; 1–2/52 → recovery but may be slow

E: **CMV** – fever, hepatitis, lymphadenopathy, rash, eye involvement (retinitis); congenital – stillbirth, growth retardation, jaundice, hepatosplenomegaly, purpura, encephalitis, microcephaly, choroidoretinitis
EBV – rash with ampicillin; malaise out of proportion to other symptoms; fever, lymphadenopathy; splenomegaly (60%); hepatomegaly (10%); jaundice (8%)

P: **EBV** – epithelial cell lysis; B-cell transformation

I: **CMV** – FBC – ↑ WBC (atypical monocytosis); LFT – mildly ↑; urine/saliva/milk/secretions/tissue – virus isolation; serology – IgG, IgM
EBV – FBC – ↑ WBC (atypical monocytosis); serology – IgM, IgG, monospot

M: **CMV** – consider ganciclovir but beware side-effects
EBV – bedrest & aspirin; treat complications; ganciclovir may be used in immunocompromised

C: **CMV** – pneumonitis; myocarditis, pericarditis; neuritis, GBS, encephalitis, meningitis, retinitis; thrombocytopaenia, haemolytic anaemia
EBV – of mononucleosis are rare – GBS, 2° bacterial infection, hepatic necrosis, thrombocytopaenia, haemolytic anaemia, splenic rupture; of EBV infection – Burkitts, Hodgkin's or T-cell lymphomas; nasopharyngeal carcinoma; oral hairy leucopaenia

P: **CMV** – death ↑ in compromised; may relapse; prevention – passive immunisation, screen blood & transplant organs
EBV – good in infectious mononucleosis

D: STI mostly affecting the lower genital tract caused by *Neisseria gonorrhoeae*

A: *N. gonorrhoeae* is a G −ve diplococcus

A/R: unprotected sex; urban areas; young; socio-economically deprived; ethnic minorities

E: worldwide; needs a sizeable population to maintain

H: 5% asymptomatic; 95% IP 2–8/7 → dysuria & discharge; ♀ pelvic infection → deep dyspareunia, lower abdominal pain

E: ♂ & ♀ mild meatitis with discharge; ♀ also abdominal tenderness

P: epithelial attachment → transmitted to lamina propria → multiplication (needs iron)

I: swab – microscopy – sensitivity in ♂ > ♀ & culture

M: penicillin or ciprofloxacin; problem of increasing resistance, contact tracing; look for other STIs

C: ♂ epididymo-orchitis; ♀ pelvic infection; facilitates acquisition & transmission of HIV; dissemination; joint involvement

P: requires repeat investigations for relapse/reinfection; prevention – use of condoms

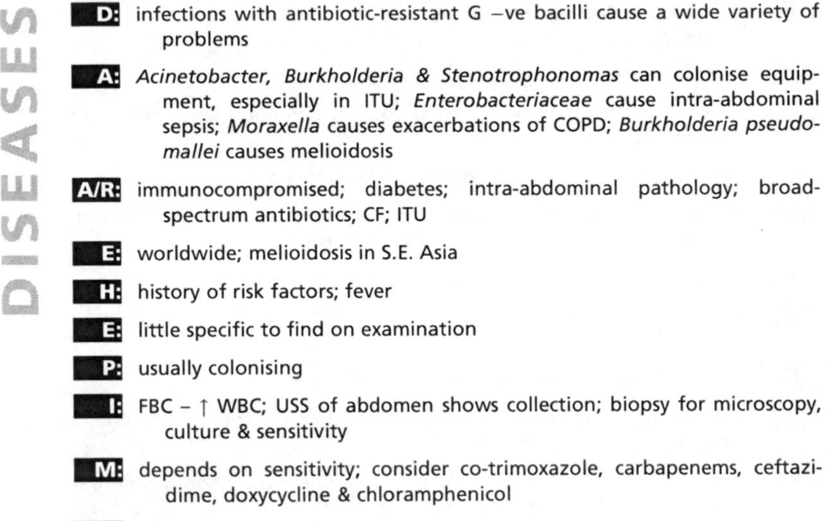

D: infections with antibiotic-resistant G −ve bacilli cause a wide variety of problems

A: *Acinetobacter, Burkholderia & Stenotrophonomas* can colonise equipment, especially in ITU; *Enterobacteriaceae* cause intra-abdominal sepsis; *Moraxella* causes exacerbations of COPD; *Burkholderia pseudomallei* causes melioidosis

A/R: immunocompromised; diabetes; intra-abdominal pathology; broad-spectrum antibiotics; CF; ITU

E: worldwide; melioidosis in S.E. Asia

H: history of risk factors; fever

E: little specific to find on examination

P: usually colonising

I: FBC – ↑ WBC; USS of abdomen shows collection; biopsy for microscopy, culture & sensitivity

M: depends on sensitivity; consider co-trimoxazole, carbapenems, ceftazidime, doxycycline & chloramphenicol

C: septicaemia

P: high mortality (up to 50%)

Haemolytic uraemic syndrome

D: syndrome of acute renal failure + thrombocytopaenia + microangiopathic haemolytic anaemia + characteristic renovascular pathology

A: most commonly due to *Escherichia coli O157* which is a G −ve bacterium; produces a verocytotoxin; transmission via contaminated animal products or faecal–oral; also due to O1014 : H21 & O111 : NM, coxsackie, *Shigella*, *Streptococcus pneumoniae* & HIV

A/R: < 5 or > 65 years of age; children ♀ = ♂ but adults ♀ > ♂; rare in Africans

E: *E. coli* widely distributed in animals; most outbreaks associated with beef products or unpasteurised milk

H: IP ??? → cramping abdominal pain, watery diarrhoea → bloody diarrhoea → HUS 6–10/7 later

E: fever; abdominal tenderness

P: toxin-mediated damage of intestinal mucosa & renal microvasculature

I: FBC – ↑ WBC, ↓ Hb; U & E – ↑ urea, ↑ creatinine, may ↑↑ urate; hyperbilirubinaemia (unconjugated); urinalysis – blood & protein; stool – microscopy, culture & sensitivity; PCR

M: treat dehydration; may need peritoneal dialysis; antibiotics controversial (may prolong or worsen illness) – ciprofloxacin in adults, cefotaxime or ceftriaxone in children

C: CNS disturbance 2° to uraemia

P: mortality up to 17% (depends on characteristics of outbreak); recovery of renal function likely in 70% survivors; some renal sequelae in up to 30%, e.g. hypertension

Haemophilus spp.

D: infections with *Haemophilus* spp. including URTI, LRTI, joint & soft tissue infection & STI chancroid

A: *Haemophilus* spp. are G −ve commensals of many animals; *H. influenzae*: *b* causes meningitis, epiglottitis, cellulitis & septic arthritis; *non-b* causes otitis, sinusitis, conjunctivitis & pneumonia; *H. ducreyi* causes chancroid

A/R: *H. influenzae*: ↓ risk if serum Abs, e.g. vaccinated or maternally derived; arthritis ↑ if pre-existing joint disease or systemic illness; < 5-year-olds; immunocompromised

H. ducreyi: ♀ carrier state; circumcised < uncircumcised; unprotected sex; immunocompromised

E: 80% of population carry *H. influenzae* but only 3–5% have type *b*; *H. ducreyi* causes > 60% of genital ulceration in Africa but rare in UK & N. America

H: *H. influenzae b*: invasive – (i) meningitis: URTI, fever, headache, vomiting, seizures; (ii) epiglottitis: sore throat, fever, dyspnoea, dysphagia, drooling; (iii) cellulitis – painful area (cheek or periorbital); (iv) arthritis – fever, joint pain (single large joint); *H influenzae b* & *non-b*: pneumonia – cough

H. ducreyi: IP 3–7/7 → painful vesicles → soft ulcers, 1–2/52 → inguinal node involvement → bubo → heals slowly but may relapse

E: *H. influenzae* – meningitis: altered CNS status, fever, neck stiffness; epiglottitis: red, swollen epiglottis (examine with great care); cellulitis: raised, warm, tender swollen area may be red/blue; arthritis: reluctance to weight bear, fever, single painful joint; pneumonia: respiratory crackles

H. ducreyi – multiple sticky haemorrhagic ulcers in genital area

P: not well understood

I: *H. influenzae* – all: FBC – ↑ WBC, blood/CSF/swab/aspirate – microscopy & culture; meningitis: CSF – ↑ WBC (polymorphs); pneumonia: CXR – consolidation

H. ducreyi – material – microscopy & culture (gold standard but difficult)

M: *H. influenzae* – all: fluid management, chloramphenicol or ceftriaxone or cefotaxime; epiglottitis: airway management +/– ventilation; arthritis: aspiration

H. ducreyi – erythromycin

C: *H. influenzae* – epiglottitis: airway obstruction, sepsis; arthritis: joint destruction

H. ducreyi: extensive local destruction especially in immunocompromised

P: *H. influenzae*: good if managed well; prevention – Hib vaccine, chemoprophylaxis for household children

H. ducreyi: prevention – promote use of condoms

D: infections with these viruses cause a haemorrhagic fever with renal syndrome

A: spread by animal hosts – Hantaan virus by *Apodemus agrarius* (striped field mouse), Puumala virus by *Clethrionomys glareolus* (bank vole) & Seoul virus by *Rattus norvegicus* (Norway rat); spread by aerosol of infectious excreta

A/R: occupational exposure – woodcutters, farmers, shepherds, military

E: Hantaan in Asia, Balklans; Puumala in Scandinavia, W. Russia, Europe; Seoul is global

H: IP generally 12–16/7 → fever, malaise, headache, myalgia, back pain, abdominal pain, nausea, vomiting, facial flushing, rash, conjunctival haemorrhage

E: fever 3–7/7 → hypotension < 3/7 → oliguria 3–7/7 → diuresis 1/7–3/52 → prolonged convalescence (may skip stages)

P:

I: FBC – ↑ WBC, ↓ platelets; U & E – ↑ urea, ↑ creatinine; ↑ LDH; ↑ APTT

M: admit & avoid trauma/movement; fluid management; dialysis if in ARF; ribavirin

C: ARF & sequelae such as hypertension

P: Hantaan mortality 5–50%; Puumala mortality < 1%; Seoul not usually fatal; prevention – no vaccine, avoid rodents

Helicobacter pylori

D: causative agent in acute & chronic non-autoimmune gastritis & PUD

A: *H. pylori* is a G −ve bacterium; produces urease; can live below mucous layer of stomach; spread by faecal–oral route; ? acquired in childhood

A/R: other factors associated with ulcer disease, e.g. race, alcohol, diet, stress

E: worldwide

H: chronic epigastric pain; nausea, vomiting, flatulence

E: epigastric tenderness

P: survives pH → penetrates mucus → attaches to epithelial cells or remains free → urease, cytotoxin & protease production → inflammation

I: non-invasive: serum/saliva – ELISA for Abs; breath test
invasive: biopsies – microscopy; culture not routine

M: triple therapy, e.g. ampicillin + metronidazole + PPI but ↑ metronidazole resistance

C: chronic diarrhoea & malnutrition in children; MALToma; complications of ulcer, e.g. perforation

P: excellent; prevention – lifestyle modification

Hepatitis A

D: viral hepatitis

A: Hepatitis A is an RNA enterovirus; transmission faecal–oral via contaminated food/water

A/R: travel to endemic area; contact with infected person; poor sanitation; overcrowding; shellfish

E: worldwide; 10 000 cases p.a. in UK (5% are imported)

H: history of exposure; IP 14–42/7 → gradual onset low-grade fever, myalgia, abdominal discomfort, anorexia, vomiting, 3–6/7 → dark urine, pale faeces, jaundice, arthralgia & rash (5%)

E: tender hepatomegaly; splenomegaly (20%); jaundice

P: direct viral cytopathicity & host immune response → hepatocellular damage

I: LFT: ↑ × 10–100 ALT or AST; ↑ APTT; ↑ bilirubin; serology for HAV IgM (+ve for 12/52) & IgM (+ve for life)

M: bedrest; vitamin K if ↑ APTT; fulminant disease may need ITU/liver transplant

C: fulminant hepatitis 0.1%; cholestatic or relapsing hepatitis; aplastic or haemolytic anaemia, post-hepatic (chronic fatigue) syndrome

P: fulminant hepatitis has 20% mortality; usually acute & mild; infection → immunity; prevention – active & passive vaccination, improved food hygiene, sanitation & water supply

Hepatitis B & D

D: viral hepatitis; may be B on its own or dual infection

A: HBV is a DNA virus; HDV is a defective RNA virus and needs HBV for replication; transmission is by blood or body fluids, or vertical (70–90% transmission rates)

A/R: IVDA; haemodialysis; receipt of blood products (especially pre-screening); male homosexuals; institutionalisation; coinfection ↑ risk of complications

E: worldwide; 200 million infections; 10 000 new in UK p.a. → lifetime risk of 5%

H: history of exposure; insidious onset fever, anorexia, upper abdominal discomfort, nausea, vomiting, distaste for cigarettes, 2–6/7 → dark urine, pale stools, jaundice

E: fever; tender abdomen, smooth tender hepatomegaly; splenomegaly (15%); jaundice

P: acute – hepatocellular damage & inflammatory infiltration; chronic – lymphocytic infiltrate in CPH and/or disturbed architecture in CAH

I: LFT – ↑ × 10–100 ALT or AST; ↑ APTT; ↑ bilirubin; serology for HBVe or s Ag or IgMcoreAb (HBVsAb in chronic); PCR for HBV +/– HDV

M: acute – bedrest; vitamin K if ↑ APTT; fulminant disease may need ITU; chronic – IFN → 35% clearance (↓ if cirrhosis, HIV, vertical, child); consider lamivudine (but resistance emerging) or tenofovir (not licensed)

C: hepatocellular failure/fulminant hepatitis (1%); carrier state (5–10%) → CPH (70%) or CAH (30%); aplastic or haemolytic anaemia; thrombocytopaenia; GBS; CFS; cholestatic & relapsing hepatitis; cirrhosis; hepatoma

P: 90% benign → complete recovery in 2–4/52; 25–30% of chronic carriers (mostly CAH) → cirrhosis +/– hepatoma; prevention – active & passive immunisation available

D: viral hepatitis

A: ssRNA virus; transmission via blood products & less frequently sexual, vertical or occupational transmission

A/R: IVDA; receipt of blood products (especially pre-screening)

E: worldwide; prevalence in UK is 1 in 1400

H: history of exposure; IP 4–26/52 → insidious onset low-grade fever, anorexia, upper abdominal discomfort, nausea, vomiting, 2–6/7 → dark urine, pale stools; may have arthralgia/arthritis

E: smooth tender hepatomegaly; splenomegaly; jaundice

P: lobular disarray; hepatocyte damage; infiltration

I: LFT – ↑ × 10 ALT or AST; ↑ APTT; ↑ bilirubin; serology for HCV Ab ELISA; PCR; HAV & HBV serology to exclude; consider biopsy

M: acute = bedrest, vitamin K if ↑ APTT; chronic = ribavirin + IFN → 70% response (depends on genotype)

C: 1–2% fulminant hepatitis; CAH; CPH; cirrhosis; hepatoma; aplastic anaemia; agranulocytosis; cryoglobulinaemia

P: 70% of HCV Ab +ve have chronic hepatitis on biopsy; up to 20% of chronics → cirrhosis in 5–30 years 15% of which → hepatoma with life expectancy < 5 years

Hepatitis E

D: viral hepatitis

A: HEV is a calicivirus; transmission is by the faecal–oral route & contaminated water

A/R: sewerage contamination of water supply

E: sporadic cases in all countries; outbreaks mainly in developing countries – S.E. Asia, Burma, former USSR, Mexico, Venezuela, N. Africa; causes 50% of non-A non-B non-C hepatitis

H: history of exposure; IP 14–42/7 → gradual onset low-grade fever, myalgia, abdominal discomfort, anorexia, vomiting, 3–6/7 → dark urine, pale faeces, jaundice, arthralgia & rash (5%)

E: tender hepatomegaly; splenomegaly (20%); jaundice

P:

I: LFT – ↑ then ↔ ALT or AST; serology for HEV IgM or IgG; PCR; HAV, HBV, HCV serology to exclude

M: bedrest; vitamin K if ↑ APTT; fulminant disease may need ITU

C: fulminant in 0.1%

P: unusually high mortality (20–40%) in pregnancy; 20% mortality from fulminant; prevention – improvements in hygiene & sanitation/water supply

D: cause a spectrum of disease – oral (cold sores), conjunctival, cutaneous, genital & encephalitis

A: HSV 1 mainly causes oral, conjunctival & cutaneous infections & encephalitis; HSV 2 mainly genital; HHV 6 roseola infantum; HHV 8 probable agent of Kaposi's sarcoma; herpes virus simiae (HVB) rare cause of encephalitis; transmission is mainly by contact with infective lesion

A/R: HSV 1 early childhood; HSV 2 unprotected sex; HHV 6 6/12–2 years of age; severe disease in HIV/AIDS or other immunocompromised; underlying skin problem

E: worldwide

H: HSV 1 & 2: IP 2–12/7 → HSV 1 febrile illness, ulcers, vesicles or encephalitis – fever, malaise, headache, nausea, vomiting; HSV 2 → fever, vesicles, ulcers, heals in 2–4/52; HHV 6: fever, rash

E: HSV 1: orofacial ulcers & vesicles or encephalitic signs including focal neurology; HSV 2: genital ulcers & vesicles; HHV 6 maculopapular rash

P: vesicles due to cell degeneration & oedema; encephalitis is acute necrotising

I: Abs in paired sera; PCR; immunofluorescence for virus

M: HSV: consider acyclovir depending on infection

C: 2° infection of skin lesions; HSV 1 & 2 latency & reactivation; HHV 6 diarrhoea, bronchopneumonia in children

P: HSV 1 encephalitis has 70% mortality if untreated; HHV 6 resolves without complication; prevention – wear gloves if occupational exposure, e.g. dentist, consider long-term acyclovir prophylaxis if immunocompromised

Histoplasmosis

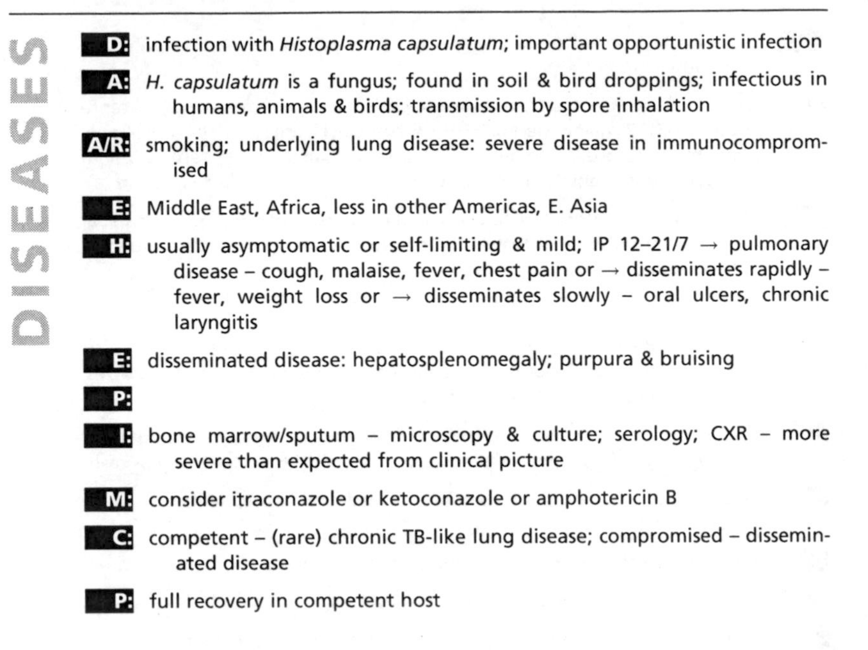

D: infection with *Histoplasma capsulatum*; important opportunistic infection

A: *H. capsulatum* is a fungus; found in soil & bird droppings; infectious in humans, animals & birds; transmission by spore inhalation

A/R: smoking; underlying lung disease: severe disease in immunocompromised

E: Middle East, Africa, less in other Americas, E. Asia

H: usually asymptomatic or self-limiting & mild; IP 12–21/7 → pulmonary disease – cough, malaise, fever, chest pain or → disseminates rapidly – fever, weight loss or → disseminates slowly – oral ulcers, chronic laryngitis

E: disseminated disease: hepatosplenomegaly; purpura & bruising

P:

I: bone marrow/sputum – microscopy & culture; serology; CXR – more severe than expected from clinical picture

M: consider itraconazole or ketoconazole or amphotericin B

C: competent – (rare) chronic TB-like lung disease; compromised – disseminated disease

P: full recovery in competent host

D: HIV infection leading to AIDS

A: HIV is a retrovirus; transmission via blood & body fluids; transmission can be sexual, vertical or via blood products; HIV 1 has genotypes A–G

A/R: IVDA; other STIs, e.g. genital ulcers; high-risk sexual behaviour; receipt of blood products (especially pre-screening)

E: HIV 1 worldwide; HIV 2 W. Africa; 42 million infected; 5 million new in 2002; 3.1 million deaths in 2002

H: history of possible exposure; seroconversion/1° → fever, rash, glandular fever-like illness, rarely CNS involvement; asymptomatic or PGL → persistent enlarged lymph nodes (painless); early disease → skin problems, e.g. seborrhoeic dermatitis; intermediate disease → persistent/ recurrent fever, diarrhoea, weight loss, candidiasis; advanced disease → AIDS

Ex: HIV infection: weight loss; fever; peripheral neuropathy, cognitive disorders; herpes, folliculitis, tinea, Kaposi's sarcoma, pruritis, dermatitis, psoriasis, thrush, gingivitis, hairy leucoplakia; lymphadenopathy; hepatosplenomegaly; chest signs; genital ulceration AIDS-defining diseases: recurrent bacterial infection in a child < 13 years of age; candidiasis of oesophagus or respiratory tract; invasive cervical cancer; extrapulmonary coccidiomycosis, cryptococcosis or histoplasmosis; cryptosporidial or isosporiasis diarrhoea > 1/12; CMV not of liver, spleen or nodes; HSV ulcer > 1/12, bronchitis, pneumonia or oesophagitis; Kaposi's sarcoma; lymphoma; disseminated mycobacteriosis or pulmonary TB; PML; PCP; recurrent pneumonia; toxoplasmosis of brain; wasting syndrome due to HIV

P: infection & critical injury of cells of the immune system including ↓ CD4+ T-cells, T-cell dysfunction & polyclonal B-cell expansion

I: all stages (except early 'window period' of 3 months) HIV Abs; CD4+ cell count – 1° ↓ but recover; PGL > 500/mm³; early > 350/mm³; intermediate ≥ 200/mm³; advanced < 200/mm³; viral load (PCR) helpful for monitoring

M: triple therapy – commonly 2 nucleoside RT inhibitors + protease inhibitor or non-nucleoside RT inhibitor, e.g. zidovudine + lamivudine + efavirenz; treat 2° problems; consider vaccinations & chemoprophylaxis; monitor CD4 count & viral load; if resistance emerges, then genotyping/phenotyping may help with future treatment planning

C: opportunistic infections; 2° cancers; CNS involvement; anaemia & thrombocytopaenia; cardiomyopathy; glomerulonephritis; adrenocortical hypofunction; pancreatitis; polyarthritis; myopathy

P: good if treated; poor if untreated; prevention – no vaccine, promote safe sex, ↓ IVDA; consider PEP; try to reduce vertical transmission with maternal treatment, Caesarean section, perinatal PEP and stopping breastfeeding

HTLV 1 & 2

D: human T-cell leukaemia/lymphoma viruses; HTLV 1 associated with ATLL & TSP; HTLV 2 associated with hairy cell leukaemia & TFP

A: both are pathogenic retroviruses; transmission by blood & body fluids, also perinatal/neonatal

A/R: IVDA; Caribbean origin; TSP ♀ > ♂ 9 : 1

E: HTLV 1 endemic in aboriginal groups, Caribbean, W. Africa, Papua New Guinea, parts of S. America but 20% cases in Kyushu, Japan

H: HTLV 1 95% asymptomatic; ATLL – non-specific symptoms, e.g. fever; TSP – progressive gradual onset difficulty walking, back & leg pain

E: ATLL: lymphadenopathy, hepatosplenomegaly; TSP: minor lower limb sensory loss, spastic paralysis

P: ATLL – transformation & malignant proliferation of T-cells; TSP – inflammation & infiltration of spinal cord

I: ATLL – FBC – ↑ WBC +/– ↓ Hb; ↑ Ca^{2+}; abnormal blood film; XR – lytic lesions; TSP – MRI; B12 levels, syphilis serology to exclude; CSF – ↑ WBC & ↑ protein

M: ATLL – refractory to chemotherapy but consider for temporary remission; palliative care; TSP – physiotherapy & OT; baclofen for spasticity

C: ATLL – 2° problems due to immunocompromise; TSP – bladder & bowel involvement

P: ATLL mean survival 6–24/12 in Japan

Influenza & parainfluenza viruses

D: viral causes of influenza, laryngotracheitis & bronchiolitis

A: influenza A, B & C are orthomyxoviruses described by type, origin, strain, year & H & N subtypes, & spread by secretions & large droplets; parainfluenza 1, 2, 3, 4A & 4B are paramyxoviruses spread by person-to-person contact, secretions & large droplets

A/R: impaired cellular immunity results in prolonged infections; attack rates are highest in children; mortality highest in the elderly, immunocompromised or those with chronic disease

E: worldwide

H: influenza – fever, chills, headache, myalgia, dry cough, rhinitis, stuffiness, sore throat; parainfluenza – fever, cough, rhinitis, sore throat

E: may look unwell or have chest signs

P: ciliated (influenza) or epithelial (parainfluenza) cell necrosis & inflammation

I: FBC – ↑ WBC

M: within 48 h influenza A & B – consider oral oseltamir or inhaled zanamavir, for A only amantidine or rimantidine, nebulised ribavirin; parainfluenza – nebulised ribavirin especially in SCID

C: influenza – 2° bacterial infection, febrile convulsions; parainfluenza – severe epithelial necrosis, parotitis, exacerbation of lung disease, febrile convulsions

P: influenza – may be fatal in high-risk patients, antigenic shift & drift used to evade host immune response; parainfluenza – infection gives partial immunity; prevention – influenza vaccination for high-risk groups

Japanese B encephalitis

D: viral cause of encephalitis

A: JBE virus is a flavivirus; amplified in pigs; transmission is via *Culex tritaeniorhynchus*

A/R: rainy season; close proximity to pigs; unvaccinated; partial cross-protective immunity with other flaviviruses

E: S., E. & S.E. Asia, Japan, Far East, Guam, former USSR, Malaysia, India, Western Pacific Islands

H: mostly asymptomatic, only 1 in 300 infections → encephalitis; IP 6–16/7 → non-specific prodrome → fever, headache, nausea, vomiting, stiff neck → tremor, ataxia, upper limb paralysis

E: altered consciousness, stiff neck, seizures, cranial nerve palsies, parkinsonism

P: neuronal degeneration & necrosis, perivascular inflammation, glial nodules in grey matter

I: CSF – capture ELISA for IgM

M: general supportive; may need an anticonvulsant

C: interstitial myocarditis; kidney haemorrhage; SIADH; spontaneous abortion; long-term neuropsychiatric disability, e.g. parkinsonism, paralysis

P: 25% mortality; prevention – vaccine, vector control

D: infection with *Legionella pneumophilia* resulting in Legionnaire's disease (pneumonia) or Pontiac fever (flu-like illness)

A: *L. pneumophilia* is a G −ve bacillus; 15 serogroups with disease mainly due to 1, 4 & 6; lives in fresh water, soil & mud, also air-conditioning systems, spas/baths, fountains, taps & showerheads; transmission is via aerosol

A/R: ♂ > ♀ 2–3 : 1; 40–70-year-olds; smoking; alcohol; diabetes; chronic illness; immunocompromised

E: worldwide

H: suggestive history; may be asymptomatic or IP 2–10/7 → Legionnaire's – fever, shivers, headache, myalgia, cough; or → Pontiac – fever, shivers, headache, myalgia, malaise, dizziness

E: Legionnaire's – looks ill/toxic, fever, chest signs, focal neurological signs/delirium/confusion (50%)

P: severe inflammatory response

I: Legionnaire's – FBC – ↑ WBC; U & E ↓Na^+, ↑ urea (50%); LFT ↑ (50%); sputum – microscopy (for pus cells) & culture; urine – Ags (serotype 1), serology – titres peak @ 2–4/52 (20% no serological response); CXR – diffuse shadowing, 25% have small pleural effusion

Pontiac – FBC, U & E, LFT normal; CXR clear; usually retrospective serology

M: Legionnaire's – erythromycin + rifampicin if deteriorating; supportive care; assisted ventilation if respiratory failure; Pontiac – supportive

C: Legionnaire's – acute respiratory failure; cardiac complications; neurological involvement; reversible acute renal failure

P: full recovery from Legionnaire's may be slow; prevention – physical/chemical water treatment, outbreak public health management

Leishmaniasis

D: visceral, cutaneous & mucocutaneous disease due to *Leishmania* spp.

A: *Leishmania* spp. are intracellular protozoa; disease most commonly due to *L. tropica, L. major, L. aethiopica, L. donovani; L. infantum, L. peruviana, L. mexicana, L. brasiliensis* & *L. chagasi*; transmission is via *Lutzomyia* & *Phlebotomus* (sandflies)

A/R: immunocompromised (especially HIV/AIDS)

E: O.W. cutaneous – *L. tropica, L. major* & *L. aethiopica* (Mediterranean, Indian subcontinent, China, SSA)

O.W. visceral – *L. donovani* (India, E. Africa), *L. infantum* (Mediterranean) & *L. chagasi* (C. & S. America)

N.W. – *L. mexicana* (Yucatan, Belize, Guatemala), *L. peruviana* (West Andes of Peru, Argentine highlands) & *L. brasiliensis* (tropical forests of S. & C. America)

H: O.W. cutaneous – IP 2–4/52 → painless papule → enlarges (3/12) → ulcer → heals over 12/12 leaving a scar

O.W. visceral – fever (2 peaks daily), sweating, malaise, weight loss, diarrhoea, cough, epistaxis

N.W. – IP 2–8/52 → cutaneous ulcerating lesion → heals or 1–12/12 (*L. brasiliensis*) → mucosal lesions (mucocutaneous disease)

E: O.W. cutaneous – circular ulcer with indurate edge & satellite lesions; regional lymphadenopathy

O.W. visceral – wasting; hepatosplenomegaly; lymphadenopathy; may be petichiae & bruising

N.W. – ulcer similar to O.W.; mucocutaneous diseases may cause destruction of nasopharynx

P: O.W. cutaneous – granulomatous response

O.W. visceral – defective CMI response

N.W. – granulomatous response

I: O.W. cutaneous – skin smear/biopsy – microscopy with Giemsa stain & culture

O.W. visceral – FBC – ↓ WBC, ↓ platelets, ↓ Hb, ↔ MCV; ↑ ESR; ↑ IgG; lymph node/bone marrow/liver/spleen biopsy – microscopy with Giemsa stain & culture; serology

N.W. – skin smear/biopsy – microscopy with Giemsa stain & culture; serology; skin test

M: O.W. cutaneous – no treatment usually needed but beware of mucocutaneous involvement

O.W. visceral & N.W. – antimony stibogluconate, pentamidine, miltefosine, +/– IFN, beware of resistance; difficult to manage in HIV

C: O.W. cutaneous – diffuse cutaneous leishmaniasis; *L. recidivans*

O.W. visceral – intercurrent infection; haemorrhage; renal amyloidosis; mucosal spread; post-kala-azar dermal leishmaniasis (up to 20%)

N.W. – aspiration pneumonia; airway obstruction

P: O.W. cutaneous – benign but may progress to chronic

O.W. visceral – good recovery with treatment; if left untreated it is often fatal within 2 years; 25% relapse rate

N.W. – worst forms are progressive, mutilating & occasionally fatal prevention – avoid sandfly bites, vector & reservoir control

D: spectrum of disease from tuberculoid to borderline to lepromatous leprosy caused by *Mycobaterium leprae*

A: *M. leprae* is an intracellular mycobacterium; transmission is via droplets/URT & broken skin

A/R: genetics

E: endemic in Africa, Indian subcontinent, S.E. Asia, C. & S. America; millions infected; important cause of morbidity & mortality in developing world

H: skin lesion → 75% heal; 25% → true leprosy – sensory skin disturbances, muscle wasting

E: solitary hypopigmented lesion +/– central sensory loss → few–many lesions with symmetrical/asymmetrical distribution; thickened nerves on palpation; altered sensation; dactylitis; lymphadenopathy; mucous membrane ulceration

P: chronic inflammation due to mycobacterial persistence

I: skin smear – microscopy with Z–N stain

M: paucibacillary rifampicin DOTS + dapsone; multibacillary – rifampicin + clofazimine both DOTS + clofazimine + dapsone also consider ethionamide, prothionamide, steroids & thalidomide for erythema nodosum leprosum (ENL) a reaction to therapy

C: blindness; bone involvement – cysts, necrosis, dactylitis; nail involvement; kidney involvement; trauma 2° to paraesthesia, ENL

P: good if treated early; chronic disability if untreated; drug reactions common; prevention – BCG offers some protection

Leptospirosis

D: zoonotic infection with *Leptospira interrogans*

A: *L. interrogans* is a spirochaete with 200 pathogenic serovars; predominant UK strains are *L. icterohaemorrhagiae* (rat reservoir) & *L. hardjo* (cow); can survive 6/12 in urine, 4/52 in fresh water & 1/7 in sea water; transmission is via infected urine damaged skin or mucous membranes

A/R: mostly ♂; occupational exposure, e.g. farmworkers; recreational exposure, e.g. kayaking

E: worldwide; mostly summer & autumn; 25–50 cases in UK p.a.

H: history of exposure; IP → high fever, rigors, headache, myalgia, abdominal pain, nausea, vomiting, cough, chest pain; may have 2nd stage → recurrence of fever, injected conjunctivae, jaundice

E: 1st stage – fever, muscle tenderness, confusion, maculopapular rash 2nd stage – Canicola fever characterised by meningism; Weil's disease characterised by jaundice; Fort Bragg fever characterised by a pretibial raised rash

P: vasculitis with injury due to immune complex deposition

I: FBC – ↑ WBC (neutrophilia) ↓ platelets (thrombocytopaenia); U & E – ↑ urea, ↑ creatinine; LFT ↑ or ↔; ↑ creatine phosphokinase; serology; Canicola – CSF – ↑ WBC; Weil's – urinalysis – proteinuria

M: benzylpenicillin IV; dialysis if in ARF

C: renal failure; myocarditis; ARDS; DIC; chronic uveitis

P: death rare; may relapse

D: infection with *Listeria monocytogenes* causing a range of disease

A: *L. monocytogenes* is a G +ve bacillus; transmission is via contaminated food

A/R: most disease in elderly & pregnant; more severe in HIV/AIDS, other immunocompromised; exposure to raw food, dairy products especially soft cheese & pate

E: worldwide

H: many infections are asymptomatic; septicaemia – history of underlying susceptibility, fever; 30–50% → meningoencephalitis – fever, stiff neck

E: septicaemia – fever, hypotension, shock; meningoencephalitis – fever, meningism, focal neurology

P:

I: FBC – ↑ WBC; ↑ CRP; blood/CSF/swabs/meconium/aspirate – microscopy & culture

M: ampicillin/amoxicillin + gentamicin (avoid in pregnancy) for 2–6/52; neonates – ampicillin/amoxicillin for > 3/52 + gentamicin for 2/52

C: amnionitis → abortion or premature labour → infected baby → long-term sequelae

P: 20–50% mortality for septicaemia & meningoencephalitis; high mortality in neonates; prevention – improved food preparation

Lyme disease

D: infection with *Borrelia burgdorferi*

A: *B. burgdorferi* is a spirochaete; transmission is by tick bite, mostly *Ixodes*

A/R: working/walking in the forests/scrub where the ticks live

E: USA, Europe, Russia, Chile, Japan, Australia

H: IP 3–32/7 → expanding annular skin lesion, fever, flu-like symptoms +/– stiff neck, headache; weeks/months → myalgia, intermittent arthritis, chest pain

E: early – skin lesion, fever; later – subacute meningitis symptoms, cranial nerve palsies, polyneuropathy; myocarditis, pericarditis

P: -

I: blood/CSF/skin – culture; PCR; serology but may be cross-reactive

M: early – tetracycline; neurological involvement – benzylpenicillin or ceftriaxone IV

C: chronic neurological, skeletal (joint erosion in 10%) or skin involvement

P: good if treated; prevention – avoid tick bites, vaccine for dogs (human vaccine withdrawn from market)

D: disease caused by infection with *Plasmodium* spp.

A: *Plasmodium* spp. are protozoa; benign malaria due to *P. malariae*, *P. ovale* & *P. vivax*; *P. falciparum* is responsible for most deaths; transmission is via the bite of the female *Anopheles* mosquito

A/R: travel to endemic region +/– poor or neglected travel advice; ↑ drug resistance; age < 5 years ↑ mortality; protective traits – sickle cell & HbF (*P. falciparum*), lack of Duffy factor (*P. vivax*)

E: *P. malariae* & *P. ovale* in Africa; *P. vivax* in Pakistan, India, Bangladesh, Sri Lanka, C. America, S.E. Asia, S. America, Oceania; *P. falciparum* and *P. vivax* Africa; 300 million cases & 3 million deaths due to *P. falciparum* p.a. (mostly children)

H: history of travel to endemic area; IP 12–28/7 depending on species → abrupt onset of fever & flu-like illness → periodicity after several days with fever, sweats, remission

E: fever; 30% have hepatosplenomegaly; jaundice

P: RBC lysis & sequestration; cytokine release

I: FBC – ↓ Hb, ↓ platelets; thick & thin blood films, Ag detection assays, e.g. OptiMal, ICT, ParaSight-F

M: see appendices

C: mostly with *P. falciparum* – cerebral malaria; severe normochromic, normocytic anaemia; tropical splenomegaly syndrome; haemolysis, renal failure & blackwater fever; pulmonary oedema; hypoglycaemia; splenic rupture (cause of death with *vivax*); shock; DIC; lactic acidosis

P: *P. malariae*, *P. ovale* & *P. vivax* have low mortality but may have recurrent infection; *P. ovale* & *P. vivax* may relapse; *P. falciparum* has mortality rates of 10–20% & 5–10% of survivors will have sequelae; prevention – bednets, insecticides, prophylaxis, vector control

Measles

D: sporadic & epidemic disease due to measles virus

A: measles virus is a paramyxovirus; it is very contagious with infection rates of 90%; transmission is via droplet

A/R: overcrowding; malnutrition; unvaccinated; immunocompromised

E: worldwide; sporadic or epidemic; relatively self-limiting in industrial nations

H: IP 10–14/7 → fever, cough, rash

E: day 3 → Koplik's spots on buccal mucosa; day 4–5 → rash on forehead/neck; day 6–9 → rash spreads to trunk

P: lysis of epithelial cells in RT & GIT; suppression of host immune system

I: clinical picture often diagnostic; serology

M: vitamin A; rehydration & nutrition; symptomatic care; treat 2° infections

C: 2° bacterial pneumonia & enteropathy; febrile convulsions, encephalitis (e.g. acute post-measles encephalitis, subacute sclerosing panencephalitis); otitis media; corneal ulceration (may cause blindness); diarrhoea/dehydration/malnutrition; sore mouth causing poor feeding; haemorrhagic measles

P: unvaccinated populations – epidemics & sporadic cases have up to 40% & 50% child mortality respectively; prevention – vaccination

D: meningitis & septicaemia due to infection with *Neisseria meningitidis*

A: *N. meningitidis* is a G –ve diplococcus; major subtypes are A, B, C, D, W135, X, Y & Z; spread by droplets; nasopharyngeal carriage 2–25%

A/R: children & young adults; overcrowding; immunocompromised

E: most common cause of bacterial meningitis worldwide; A causes epidemics in W. & E. Africa, Middle East, Nepal, India; B then C most important in western countries

H: IP 1–3/7 → fever, headache, restlessness, stiff neck, rash (60%) +/– drowsiness

E: meningitis – fever, irritability, rash (60%), neck stiffness, +ve Kernig's sign
septicaemia – fever, looks toxic, shock, rash (60%), signs of DIC

P: systemic invasion +/– meningeal inflammation; DIC

I: FBC – ↑ WBC; U & E may ↑ urea, ↑ creatinine; CSF – turbid, ↑↑ neutrophils, ↑ protein, ↓ glucose; blood/CSF – microscopy, culture & sensitivity

M: monitoring; airway management & O_2; fluid balance; pain relief; local protocol or benzylpenicillin; consider dexamethasone to ↓ complications

C: Waterhouse–Friderichsen syndrome (acute adrenal failure); hydrocephalus, CNS damage, subdural bleeds, abscess, deafness, SIADH; arthritis; vasculitis; pericarditis

P: overall mortality 5–10%, mostly from septicaemia; prevention – vaccine (none available for B), chemoprophylaxis for contacts

Molluscum contagiosum

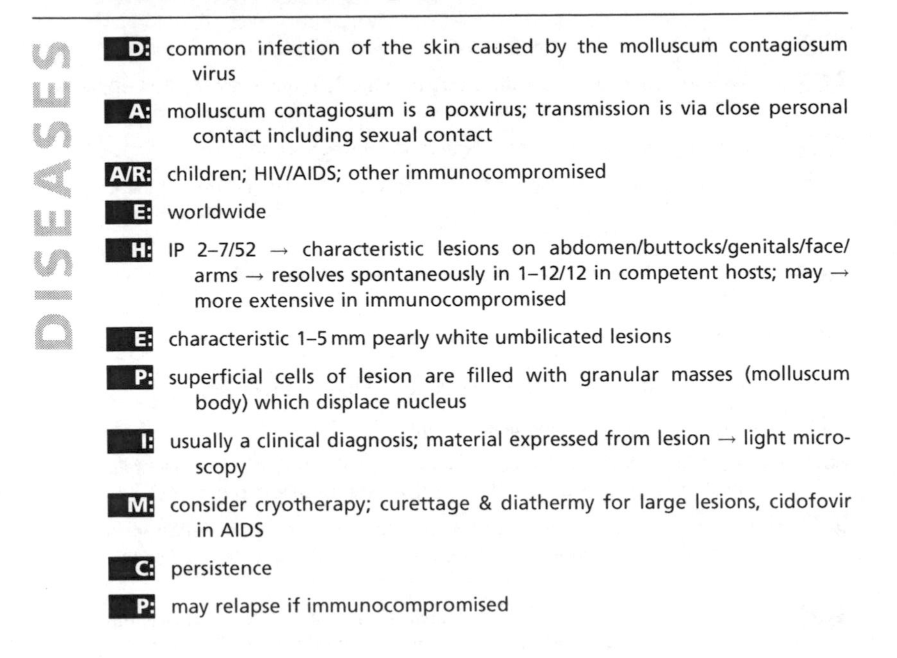

D: common infection of the skin caused by the molluscum contagiosum virus

A: molluscum contagiosum is a poxvirus; transmission is via close personal contact including sexual contact

A/R: children; HIV/AIDS; other immunocompromised

E: worldwide

H: IP 2–7/52 → characteristic lesions on abdomen/buttocks/genitals/face/arms → resolves spontaneously in 1–12/12 in competent hosts; may → more extensive in immunocompromised

E: characteristic 1–5 mm pearly white umbilicated lesions

P: superficial cells of lesion are filled with granular masses (molluscum body) which displace nucleus

I: usually a clinical diagnosis; material expressed from lesion → light microscopy

M: consider cryotherapy; curettage & diathermy for large lesions, cidofovir in AIDS

C: persistence

P: may relapse if immunocompromised

D: disease affecting the RT, GIT & skin due to mucor-like zygomycetes

A: most common fungi involved are *Absidia* spp., *Rhizopus* spp. & *Rhizomucor* spp.

A/R: diabetic ketoacidosis (rhinocerebral disease); leukaemia & immunocompromised (lung & disseminated disease); malnutrition (GI disease); burns & wounds (cutaneous disease)

E: worldwide; rare

H: rhinocerebral – fever, unilateral facial pain, black nasal discharge
GI – diarrhoea, melaena, haematemesis
pulmonary – cough, chest pain

E: rhinocerebral – facial swelling, proptosis
GI – peritonitis (due to perforation), melaena, haematemesis
pulmonary – chest signs

P: blood vessel invasion → thrombosis → infarction & necrosis

I: diagnosis mainly clinical – infection & infarction; FBC – ↓ neutrophils; culture often difficult; histology of specimens; CXR, AXR, sinus series

M: surgical debridement; amphotericin B IV or local instillation, e.g. sinuses

C: orbital invasion may cause blindness; dissemination; GI → perforation, haemorrhage

P: almost invariably fatal once spread beyond 1° site

Mumps

D: generally mild acute infection due to mumps virus

A: mumps virus is a paramyxovirus; spread by droplet or salivary contact

A/R: schoolchildren & young adults

E: worldwide; common; cyclical & seasonal peaks

H: IP 12–25/7 → prodrome of headache, sore throat, malaise, fever → salivary gland inflammation with jaw pain, earache → resolves in 1–2/52

E: 70% have parotitis, unilateral then bilateral

P: virus replicates in epithelium of URT then disseminates

I: diagnosis mainly clinical; FBC – ↔ or ↓ WBC; ↑ serum amylase; saliva/CSF/ urine – virus isolation; 4× ↑ or ↓ or single high Ab titre

M: symptomatic treatment – mild analgesics, mouthwash, soft diet; orchitis – pain relief, ice packs, support

C: meningitis, encephalitis, deafness, facial palsy, myelitis, GBS; epididymo-orchitis, oophoritis; pancreatitis; arthritis; myocarditis, pericarditis

P: death very rare; sterility & deafness rare; prevention – vaccine

D: cause a spectrum of disease from mild inapparent infection to severe pneumonia

A: *Mycoplasma* spp. are the smallest free-living microorganisms; 14 species constitute normal or pathological flora of humans, mostly in the oropharynx; disease mostly due to *M. hominis* (see bacterial vaginosis) & *M. pneumoniae* but also *M. genitalium* & *Ureaplasma urealyticum*; transmission is slow & by droplets from person-to-person contact

A/R: 5–15-year-olds; severe disease in immunocompromised

E: worldwide; endemic & epidemic; 5% community-acquired pneumonias in UK

H: may be asymptomatic; IP 6–23/7 → malaise, headache; 1–5/7 → cough (33% productive), diarrhoea & vomiting (15–25%), arthritis (15–25%); variable course, often protracted

E: may have unilateral or bilateral chest signs; mild skin rash (15–25%); signs of arthritis (15–25%)

P: adherence to respiratory epithelium; partially immune-mediated pathology

I: LFT – mildly deranged; sputum/tissue – microscopy, culture & sensitivity; CXR – more features than expected from examination, 80% show consolidation, 30% bilaterally; serology – ↑ ×4 or single high titre on ELISA; PCR; cold haemagglutins (50%)

M: tetracycline or erythromycin

C: rare but include respiratory failure, ARDS, pleural effusion (15%), cavitation, pneumocoele, bronchiectasis, fibrosis; haemolytic anaemia; CNS involvement; hepatitis, pancreatitis; myocarditis, pericarditis; glomerulonephritis

P: rarely fatal; may relapse; Abs protective

Nocardiosis

D: systemic or cutaneous disease due to *Nocardia* spp.

A: *Nocardia* spp. are G +ve bacteria; mostly due to *N. asteroides* but also *N. brasiliensis*, *N. caviae* & *N. transvalensis*; live in soil especially decaying vegetable matter; transmission is by inhalation

A/R: malignancy; HIV/AIDS; transplantation; steroids; diabetes; other immunocompromised; pre-existing pulmonary disease; rare in childhood

E: worldwide

H: infection may be asymptomatic; pulmonary – chronic cough, fever; disseminated – stuff neck

E: pulmonary – chest signs; disseminated – signs of space-occupying lesion

P: abscess formation

I: FBC – ↑ WBC; pus/BAL/tissue samples – microscopy, modified ZN stain & culture; serology @ reference centre; CXR – infiltration, cavitation, nodules

M: high-dose sulphonamide, e.g. co-trimoxazole, ampicillin or minocycline; carbapenem + 3rd generation cephalosporin + amikacin for brain abscess

C: dissemination in compromised hosts → meninges, also skin, liver, kidneys & bone

P: disseminated & CNS forms have poor prognosis in compromised hosts

D: infection with human papillomaviruses produces a wide range of diseases including anogenital warts, cervical cancer, respiratory papillomatosis, oral & skin warts

A: HPV are papovaviridae; anogenital & vulval warts & respiratory papillomatosis are caused by HPV 6, 11 & 16; oral warts – HPV 6, 11, 13 & 32; cervical cancer & CIN – 16 & 18; skin warts – HPV 1, 2, 3, 4 & 10; nasal papilloma – HPV 59; conjunctival disease – HPV 6 + 11; transmission is via direct contact or contaminated surfaces

A/R: multiple sexual partners; other STIs; being born to an infected mother (respiratory papillomatosis)

E: worldwide

H: long IP → warty growths, many regress spontaneously

E: range of characteristic lesions including painless nodules, papular warts, flat planar warts, fleshy cauliflower like lesions, cervical lesions

P: epidermal hyperplasia

I: mainly a clinical diagnosis; samples – immunofluorescence or histology

M: cryotherapy or surgery; local application of an antimitotic agent

C: CIN & cervical cancer

P: excellent

Parvovirus B19

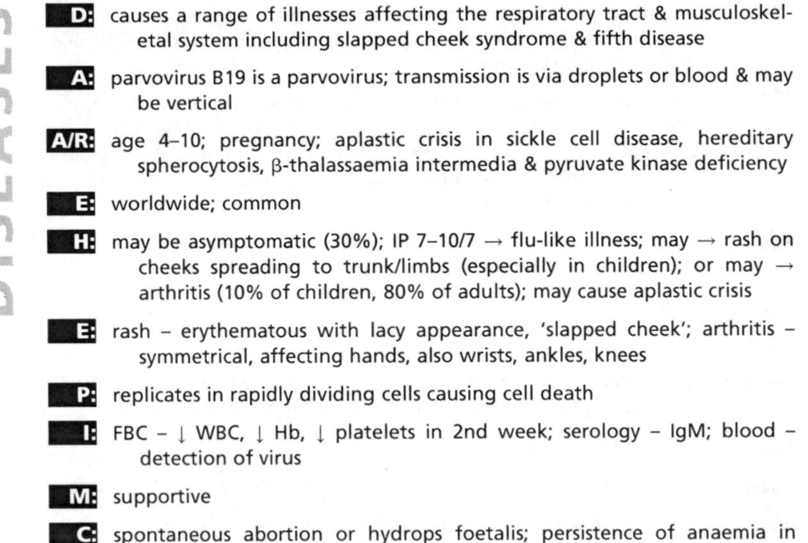

D: causes a range of illnesses affecting the respiratory tract & musculoskeletal system including slapped cheek syndrome & fifth disease

A: parvovirus B19 is a parvovirus; transmission is via droplets or blood & may be vertical

A/R: age 4–10; pregnancy; aplastic crisis in sickle cell disease, hereditary spherocytosis, β-thalassaemia intermedia & pyruvate kinase deficiency

E: worldwide; common

H: may be asymptomatic (30%); IP 7–10/7 → flu-like illness; may → rash on cheeks spreading to trunk/limbs (especially in children); or may → arthritis (10% of children, 80% of adults); may cause aplastic crisis

E: rash – erythematous with lacy appearance, 'slapped cheek'; arthritis – symmetrical, affecting hands, also wrists, ankles, knees

P: replicates in rapidly dividing cells causing cell death

I: FBC – ↓ WBC, ↓ Hb, ↓ platelets in 2nd week; serology – IgM; blood – detection of virus

M: supportive

C: spontaneous abortion or hydrops foetalis; persistence of anaemia in compromised hosts

P: prevention – no active vaccine, but Ig available

D: infections with *Pasteurella* spp. causes wound infections

A: *Pasteurella* spp. are G −ve bacilli; disease mostly due to *Pasteurella multocida*; transmission is via animal bites

A/R: domestic animals

E: worldwide

H: dog or cat bite

E: skin or soft tissue infection; signs of bone/joint involvement

P: suppurative

I: blood/pus – culture & sensitivity

M: co-amoxiclav; surgical drainage

C: septicaemia

P: good if treated & competent host

Plague

D: zoonotic disease caused by *Yersinia pestis*; 90% of cases are bubonic

A: *Y. pestis* is a G −ve coccobacillus; bubonic plague is transmitted mainly by the bites of fleas that live on *Rattus rattus* & *Rattus norvegicus*; pneumonic plague is transmitted by droplets

A/R: contact with rats; travel to endemic area

E: consistently reported from Africa, Asia & Americas; seasonal pattern with most cases in warm, dry periods

H: exposure to rats; travel to endemic area; IP 1–7/7 bubonic plague → fever, headache, painful buboes/lymph nodes; or pneumonic plague → headache, malaise, fever, vomiting, ↓ consciousness → cough, dyspnoea, haemoptysis

E: bubonic → fever, lymphadenopathy, hypotension, hepatomegaly, buboes (especially in groin); pneumonic → chest signs, signs of respiratory failure

P: congestion & necrosis

I: FBC – ↑↑ WBC, ↓ or ↔ platelets; blood/sputum/CSF/aspirate – microscopy, culture & sensitivity; serology; CXR – consolidation

M: supportive bubonic – tetracycline or doxycycline pneumonic – gentamicin or streptomycin meningeal involvement – chloramphenicol

C: bacteraemia & sepsis; pharyngeal or tonsillar abscess; meningitis; pneumonia

P: pneumonic plague is rapidly fatal; prevention – vaccine, vector & reservoir control

Pneumocystis carinii

D: opportunistic infection that causes disease in compromised hosts

A: *P. carinii* is a fungus (previously thought to be a protozoa); transmission is via inhalation

A/R: CD4+ T-cell $< 200\ mm^3$

E: worldwide; now the most common HIV/AIDS-diagnosing condition in western countries

H: asymptomatic in competent hosts; in compromised hosts → insidious onset fever, non-productive cough, ↑ SOB

E: chest signs in advanced disease

P: irreversible fibrotic reaction

I: sputum/BAL – microscopy, culture & sensitivity; CXR – bilateral diffuse interstitial infiltration +/– cavities

M: co-trimoxazole or pentamidine + prednisolone if cyanosed

C: pneumothorax due to pneumocoele formation

P: good outcomes with prompt treatment of 1st attack; relapse or failure of 1st line treatment has a poor prognosis; prevention – chemoprophylaxis with co-trimoxazole, dapsone or pentamidine, try and correct CD4 count, prophylaxis can be stopped when CD4 count > 200 cells/ml

Poliomyelitis

DISEASES

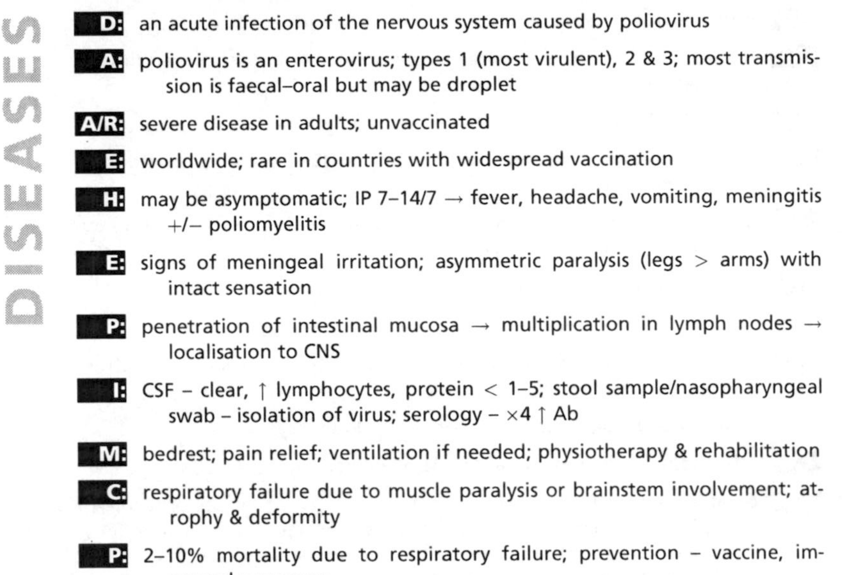

D: an acute infection of the nervous system caused by poliovirus

A: poliovirus is an enterovirus; types 1 (most virulent), 2 & 3; most transmission is faecal–oral but may be droplet

A/R: severe disease in adults; unvaccinated

E: worldwide; rare in countries with widespread vaccination

H: may be asymptomatic; IP 7–14/7 → fever, headache, vomiting, meningitis +/– poliomyelitis

E: signs of meningeal irritation; asymmetric paralysis (legs > arms) with intact sensation

P: penetration of intestinal mucosa → multiplication in lymph nodes → localisation to CNS

I: CSF – clear, ↑ lymphocytes, protein < 1–5; stool sample/nasopharyngeal swab – isolation of virus; serology – ×4 ↑ Ab

M: bedrest; pain relief; ventilation if needed; physiotherapy & rehabilitation

C: respiratory failure due to muscle paralysis or brainstem involvement; atrophy & deformity

P: 2–10% mortality due to respiratory failure; prevention – vaccine, improved sewerage

D: polyomaviruses cause progressive multifocal leucoencephalopathy

A: JC & BK are papovaviridae; the mode of their transmission is unknown but they are common in the general population

A/R: pregnancy; AIDS; other immunocompromised

E: widely distributed; disease is due to viral reactivation

H: insidious onset of speech & visual disturbance in compromised patients

E: signs of multifocal brain disease – impaired speech & vision, mental deterioration

P: remains latent in kidney; reactivation causes CNS localisation with focal patches of demyelination & necrosis

I: CT or MRI – lesions; biopsy or CSF – PCR

M: supportive; try to reverse immunocompromise

C: neurological sequelae

P: relentless progress resulting in death within 3–6/12; may rarely stabilise

Prions

D: infectious agents that target the CNS causing spongiform encephalopathy ('slow virus diseases', transmissible dementias) – CJD, nvCJD & kuru

A: prions are free nucleic acids; affect humans & animals; they are resistant to common sterilisation procedures

A/R: CJD – 45–75-year-olds, treatment with human-derived GH, exposure to infected material; nvCJD – eating contaminated meat; kuru – cannibalism

E: CJD – worldwide with an incidence of 1 per million; nvCJD – Europe; kuru – cannibalistic tribes of Papua New Guinea

H: CJD – unknown IP → prodromal fatigue, depression, weight loss, headache, malaise → multifocal dementia → akinetic mutism
kuru – history of cannibalism, ambulatory ataxia → sedentary ataxia

E: CJD → dementia, myoclonus, ataxia, pyramidal & extrapyramidal signs
kuru → cerebellar ataxia → myoclonus & spasticity

P: neuronal degeneration & loss; lack of inflammatory response

I: EEG – pseudoperiodic sharp wave activity; other tests normal but useful to exclude other causes

M: supportive; no current treatment affects outcome

C: death

P: relentlessly progressive & fatal; prevention – use of disposable surgical instruments, better abattoir practice

D: mainly opportunistic infection of a host that is compromised in some way

A: *Pseudomonas aeruginosa* is a G −ve bacterium

A/R: CF; severe burns; neutropaenia; contact lens wear (corneal keratitis); IVDA (endocarditis & osteomyelitis); diabetes (malignant otitis externa); swimming (otitis externa)

E: worldwide

H: history of predisposition;
pulmonary – productive cough
otitis externa – painful, itchy, discharging ear
corneal keratitis – very painful, red eye
osteomyelitis – painful limb/joint, fever
endocarditis – general malaise, fever

E: pulmonary – productive cough
otitis externa – inflamed ear canal, discharge
corneal keratitis – signs of keratitis/ulceration on staining & slit lamp investigation
osteomyelitis – painful limb/joint, red & inflamed area
endocarditis – may have a murmur or signs of septic emboli

P: colonisation

I: sputum/swab/blood – microscopy, culture & sensitivity, mucoid colony formation

M: antipseudomonal penicillin or carbapenem + aminoglycoside in neutropaenia or endocarditis; eye/ear drops for keratitis/otitis or oral ciprofloxacin

C: often the infection that proves fatal in CF

P: depends on underlying pathology

Q fever

D: zoonotic infection with *Coxiella burnetii* causes pneumonia & endocarditis

A: *C. burnetii* is a rickettsia-like bacterium; transmission is by aerosol or contaminated milk

A/R: occupational exposure, e.g. vets, abattoir work; unpasteurised milk

E: cause of < 1% of community-acquired pneumonia; cause of culture-negative endocarditis

H: history of work with animal; IP 2–4/52 → sudden onset fever, headache, myalgia, cough

E: hepatosplenomegaly (50%); chest signs (15%)

P: granulomatous response, especially in liver

I: serology – × 4 ↑ over 2/52 or single high Ab IgM titre; CXR – evidence of pneumonia (10–30%)

M: tetracycline or erythromycin, rifampicin or ciprofloxacin; endocarditis requires months of combination therapy, e.g. rifampicin + tetracycline

C: endocarditis (10%); hepatitis or cirrhosis

P: endocarditis has 15% mortality; otherwise fatalities rare

D: acute CNS infection with rabies virus

A: rabies virus is a rhabdovirus, part of the lyssavirus genera; inactivated by heat; transmission is by the bite of an infected animal or inoculation of mucous membranes

A/R: unvaccinated; exposure to dogs & bats

E: widespread worldwide; can cross borders in animal hosts; endemic in tropics; 0.01% of disease is in temperate areas; UK, Scandinavia, Spain, Portugal & Australia are currently free of the disease

H: history of animal bite; IP 20–90/7 → prodrome with itching, pain or parasthaesia @ site of bite → fever, chills, malaise, weakness, headache

E: neuropsychiatric symptoms; flaccid paralysis (spinal involvement); furious rabies (brain involvement) → hydrophobia, hyperaesthesia, ↑ arousal, cranial nerve defects, meningism, ANS changes

P: cerebral congestion; Negri bodies in large neurons of hippocampus & Purkinje cells of cerebellum & medulla

I: saliva/throat swab/tracheal swab/eye swab/CSF (1st week) – isolation of virus; hairy skin biopsy from nape of neck – FAT for Ag; serology in unvaccinated patients

M: if even minor suspicion give PEP – wound treatment, active & passive immunisation (with Ig); observe animal if possible

C: focal degradation of salivary glands, liver, pancreas, adrenals & lymph nodes; aspiration pneumonia, bronchitis, pneumonitis, pneumothorax; myocarditis, arrythmias; haematemesis

P: almost always fatal once symptoms have begun; correct post-exposure treatment results in a reduction to < 1% fatalities; prevention – vaccine for humans & animals

Rat-bite fevers

D: infections due to *Spirillum minus* & *Streptobacillus moniliformis*

A: *S. minus* is a G −ve spirillum that causes sodoku & is transmitted from rats or their predators via bites or scratches; *S. moniliformis* causes Haverhill fever & transmission is via rat bites or milk contaminated with rat urine

A/R: close proximity to rats; < 6-year-olds; sleeping; neuropathy

E: both worldwide; sodoku particularly in Japan

H: sodoku – history of exposure to rats; bite heals; IP 5–30/7 → fever, rigors, myalgia, prostration, arthralgia; wound may break down again
Haverhill fever – history of exposure to rats; IP 1–10/7 → high fever, vomiting, severe headache, myalgia, arthritis/arthralgia (50%)

E: sodoku → lymphadenopathy, exanthema spreading from bite
Haverhill fever → muscle tenderness, fever, evidence of rat bite, erythematous macular rash, lymphadenopathy, arthritis/arthralgia is asymmetrical, polyarticular & migrational

P: local inflammation +/− necrosis that spreads to lymph nodes

I: sodoku – FBC – ↑ WBC; aspirate – microscopy with Giemsa or Wright's stain
Haverhill – FBC – ↑ WBC; blood/joint aspirate/pus – culture

M: both – procaine benzylpenicillin or penicillin V

C: sodoku – meningitis, encephalitis; abscess formation; endocarditis, myocarditis; pleural effusion; chorioamnionitis; liver & kidney involvement
Haverhill fever – bronchitis, pneumonia; abscess formation; myocarditis, pericarditis, endocarditis; glomerulonephritis; splenitis; amnionitis; anaemia

P: sodoku – untreated has a mortality of 2–10%; survivors may relapse
Haverhill fever – untreated has a mortality of 10–15%; may have persistent arthralgia

D: viral pathogen that causes U & LRTI infections

A: RSV is a paramyxovirus; subgroups A (or 1) & B (or 2); transmission is via large droplets

A/R: peak incidence @ 2–6/12 of age; exposure to other children; premature babies; bottle-feeding; impaired T-cell immunity; lung disease; passive tobacco smoke

E: worldwide; temperate climates → epidemics every winter for 4/12; tropics → epidemics in the rainy season

H: IP 2–8/7 → wheeze, cough, rhinorrhoea, fever

E: wheeze

P: epithelial necrosis; host response may contribute to pathology

I: nasopharyngeal secretions – rapid Ag detection; pulse oximetry for O_2 saturation

M: supportive; O_2 if saturation low; consider RSV Ig or palivizumab; no evidence for ribavirin

C: may trigger an asthma attack

P: infections tend to become progressively milder; prevention – cohorting, handwashing

Roundworms (intestinal)

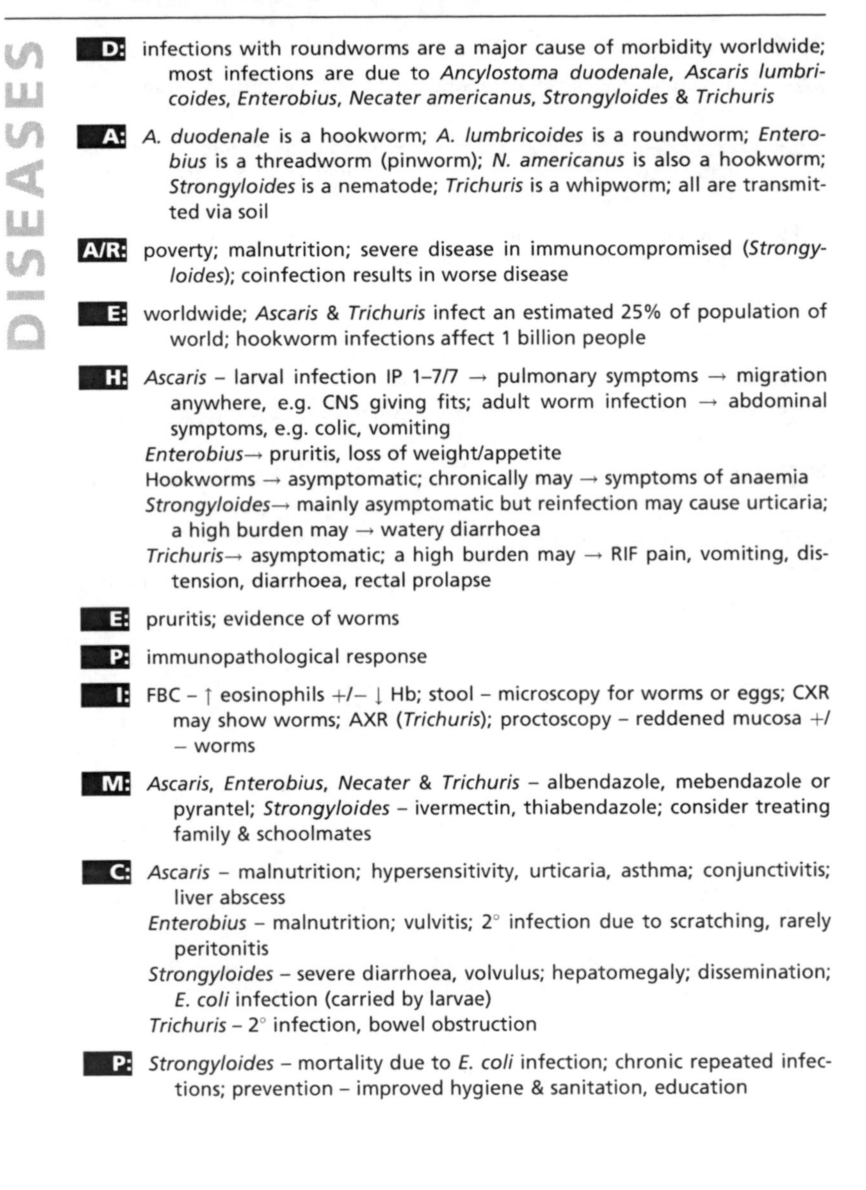

D: infections with roundworms are a major cause of morbidity worldwide; most infections are due to *Ancylostoma duodenale*, *Ascaris lumbricoides*, *Enterobius*, *Necater americanus*, *Strongyloides* & *Trichuris*

A: *A. duodenale* is a hookworm; *A. lumbricoides* is a roundworm; *Enterobius* is a threadworm (pinworm); *N. americanus* is also a hookworm; *Strongyloides* is a nematode; *Trichuris* is a whipworm; all are transmitted via soil

A/R: poverty; malnutrition; severe disease in immunocompromised (*Strongyloides*); coinfection results in worse disease

E: worldwide; *Ascaris* & *Trichuris* infect an estimated 25% of population of world; hookworm infections affect 1 billion people

H: *Ascaris* – larval infection IP 1–7/7 → pulmonary symptoms → migration anywhere, e.g. CNS giving fits; adult worm infection → abdominal symptoms, e.g. colic, vomiting

Enterobius→ pruritis, loss of weight/appetite

Hookworms → asymptomatic; chronically may → symptoms of anaemia

Strongyloides→ mainly asymptomatic but reinfection may cause urticaria; a high burden may → watery diarrhoea

Trichuris→ asymptomatic; a high burden may → RIF pain, vomiting, distension, diarrhoea, rectal prolapse

E: pruritis; evidence of worms

P: immunopathological response

I: FBC – ↑ eosinophils +/– ↓ Hb; stool – microscopy for worms or eggs; CXR may show worms; AXR (*Trichuris*); proctoscopy – reddened mucosa +/– worms

M: *Ascaris*, *Enterobius*, *Necater* & *Trichuris* – albendazole, mebendazole or pyrantel; *Strongyloides* – ivermectin, thiabendazole; consider treating family & schoolmates

C: *Ascaris* – malnutrition; hypersensitivity, urticaria, asthma; conjunctivitis; liver abscess

Enterobius – malnutrition; vulvitis; 2° infection due to scratching, rarely peritonitis

Strongyloides – severe diarrhoea, volvulus; hepatomegaly; dissemination; *E. coli* infection (carried by larvae)

Trichuris – 2° infection, bowel obstruction

P: *Strongyloides* – mortality due to *E. coli* infection; chronic repeated infections; prevention – improved hygiene & sanitation, education

Rubella

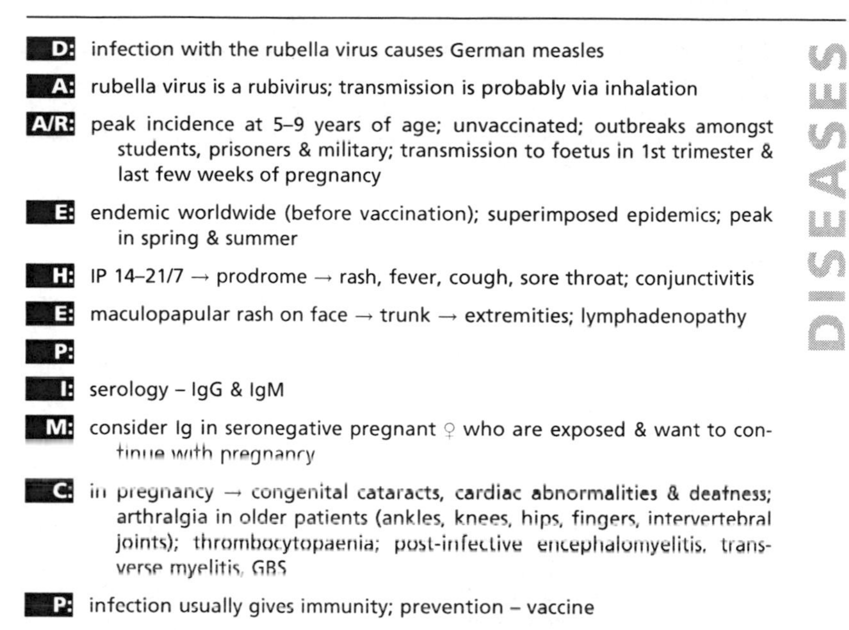

D: infection with the rubella virus causes German measles

A: rubella virus is a rubivirus; transmission is probably via inhalation

A/R: peak incidence at 5–9 years of age; unvaccinated; outbreaks amongst students, prisoners & military; transmission to foetus in 1st trimester & last few weeks of pregnancy

E: endemic worldwide (before vaccination); superimposed epidemics; peak in spring & summer

H: IP 14–21/7 → prodrome → rash, fever, cough, sore throat; conjunctivitis

E: maculopapular rash on face → trunk → extremities; lymphadenopathy

P:

I: serology – IgG & IgM

M: consider Ig in seronegative pregnant ♀ who are exposed & want to continue with pregnancy

C: in pregnancy → congenital cataracts, cardiac abnormalities & deafness; arthralgia in older patients (ankles, knees, hips, fingers, intervertebral joints); thrombocytopaenia; post-infective encephalomyelitis, transverse myelitis, GBS

P: infection usually gives immunity; prevention – vaccine

Salmonellosis (non-typhoid)

D: infection due to enterotoxin-producing non-typhoid *Salmonella* spp.

A: *Salmonella* spp. are G –ve bacilli; 2000 serotypes; most non-typhoid diseases are caused by *S. agona*, *S. enteritidis*, *S. heidelberg*, *S. indiana*, *S. typhimurium* & *S. virchow*; transmission is via contaminated food or drink from an animal source

A/R: living in an institution; severe in elderly & immunocompromised (especially HIV/AIDS); pet turtles (*S. arizonae*)

E: worldwide; more common in developed countries

H: IP 12–48/7 → nausea, vomiting, malaise, headache, fever → cramping abdominal pain, diarrhoea → settles in a few days

E: production of large volume watery stools → small volume bloody stools

P: enterotoxin causes transport defects in small intestine & inflammation in colon/lower ileum

I: FBC – ↑ WBC; U & E – ↑ urea; stool – culture, microscopy & sensitivity; blood cultures if very ill

M: oral rehydration; ciprofloxacin or cefotaxime or ceftriaxone in severe or invasive disease

C: severe dehydration; renal failure; colitis, ileitis, post-infectious IBS; reactive arthritis; invasive disease

P: rarely fatal; prevention – improved food preparation & hygiene

Schistosomiasis

D: chronically debilitating & potentially fatal infection caused by *Schistosoma* spp.; also known as bilharzia

A: *Schistosoma* spp. are blood flukes; major infections in humans are due to *S. haematobium*, *S. japonicum* & *S. mansoni*; transmission is via cercarial penetration of intact skin from fresh water where the snail host lives

A/R: travel to endemic area; exposure to snail habitat

E: 600 million @ risk & 200 million infected worldwide; *S. haematobium* is found in Africa & Middle East, & causes urinary disease; *S. japonicum* in the Far East, & causes intestinal disease; *S. mansoni* in Africa, Middle East & S. America, & causes intestinal disease

H: history of travel or exposure; @ infection may cause swimmer's itch; acute infection (Katayama fever) → fever, urticaria, headache, abdominal pain, diarrhoea; chronic intestinal or hepatic infection → occasional bloody stools or chronic urinary infection → dysuria, frequency, haematuria

E: chronic intestinal or hepatic infection → hepatosplenomegaly

P: granuloma formation @ site of egg deposition in mesenteric or bladder veins; repeated infection needed to ↑ worm load

I: FBC – ↑ eosinophils; stool/urine/biopsy – microscopy for eggs; serology for ↑ IgG or presence of IgM or IgA

M: praziquantel

C: bladder cancer; pulmonary fibrosis & hypertension; colorectal cancer; hepatitis & cirrhosis

P: need repeated infections to produce cancer; prevention – sanitation, education, ↓ snail host, ↓ water contact

Scrub typhus

D: acute febrile illness due to *Orientalis tsutsugamushi*

A: *O. tsutsugamushi* is a rickettsia; six serotype; transmission is by mite bite

A/R: occupational – oil palm & rubber estate workers, police & soldiers

E: rural Asia

H: history of exposure & bites; IP 5–10/7 → febrile illness, painful lymph nodes, drowsiness, apathy, headache, nausea, vomiting, tinnitus, hyperacusis, constipation, epistaxis, dry cough, rash

E: eschar (50–80%); axillary or groin lymphadenopathy; hepatosplenomegaly; severe disease may also show signs of meningoencephalitis

P: vasculitis & perivasculitis of small blood vessels

I: diagnosis commonly made on history & signs; serology; PCR

M: supportive care; doxycycline

C: meningoencephalitis; myocarditis

P: immunity is strain-specific & only lasts a few months; prevention – suitable clothing, repellant; avoidance

Shigellosis

D: dysentery due to infection with *Shigella* spp.

A: *Shigella* spp. are G –ve bacteria; 4 subgroups/subspecies – *S. dysenteriae* (A), *S. flexneri* (B), *S. boydii* (C) & *S. sonnei* (D); transmission is via the faecal–oral route or contaminated flies

A/R: children

E: worldwide; serious disease in developing countries is due to *S. dysenteriae* & *S. flexneri*; *S. sonnei* is endemic in developed countries

H: IP 2–4 days → headache, fever, abdominal discomfort → watery diarrhoea +/– blood; *S. dysenteriae* & *S. flexneri*→ blood & mucus; *S. sonnei*→ stays watery; severe disease → dysentery with abdominal cramps, tenesmus, small volume 'stools' made up of blood, pus & mucus

E: abdominal tenderness

P: inflammation, ulceration, haemorrhage & sloughing

I: FBC – ↑ WBC; stool – microscopy, culture & sensitivity

M: rehydration; severe cases ciprofloxacin (adults) or nalidixic acid (children)

C: toxic dilatation +/– perforation; HUS with *S. dysenteriae*; rare with *S. sonnei*

P: good with adequate rehydration; prevention – improved hygiene & sanitation

Smallpox

D: disease caused by infection with variola virus; may be used as a biological weapon

A: variola virus is a poxvirus; transmission is via inhalation or inoculation

A/R: extremes of age; pregnancy; immunocompromised

E: 'eradicated' due to vaccination campaign – certified in 1979

H: IP 12/7 → headache, fever, malaise, vomiting, rash

E: characteristic rash with centrifugal papules → pustules → heals in 2–3/52

P:

I: isolate virus from lesions, EM

M: isolation; ? role for cidofovir, vaccination for contacts

C: toxaemia; eye involvement; scarring

P: high mortality in haemorrhagic or flat, confluent types, vaccination

D: infection with *Sporothrix schenckii* causing cutaneous or deep mycosis

A: *S. schenckii* is a dimorphic fungus; lives in soil & plant matter; transmission is by inoculation

A/R: HIV/AIDS; other immunocompromised; diabetes; alcohol abuse; occupational, e.g. florists, packers, fishermen, armadillo hunters

E: widely distributed in tropics; usually sporadic but small outbreaks may occur

H: solitary ulcerated lesion on exposed site; or lymphangitic with 2° lesions; or disseminated disease

E: solitary lesion +/− small satellites; 2° lesions along path of lymphatics; dissemination to joints, lungs or meninges gives corresponding signs

P: granuloma formation

I: swab/tissue samples microscopy, culture & sensitivity; biopsy shows granulomatous response; skin test

M: potassium iodide or itraconazole or terbinafine

C: dissemination in AIDS

P: good in immunocompetent

Spotted fevers

D: infections caused by *Rickettsia* spp.

A: *Rickettsia* spp. are rickettsiae; Israeli tick typhus is caused by *R. sharoni*, Japanese tick typhus by *R. japonica*, murine tick typhus by *R. typhi*, Queensland tick typhus by *R. australis*, RMSF tick typhus by *R. rickettsii*, Siberian tick typhus by *R. sibirica* & S. African tick typhus by *R. conorii*; louse borne by *R. prowazekii*; transmission is via tick bites or inoculation of tick material by other hosts

A/R: exposure to ticks

E: *R. conorii* is found in Africa, India, Mediterranean & Middle East; *R. rickettsii* in USA, Canada & S. America; *R. sharoni* in Israel; *R. sibirica* in E. Europe; *R. typhi* is found on every continent except Antarctica

H: history of exposure to ticks; IP 1–2/52 → fever, headache, myalgia, dry cough +/– eschar → rash

E: rash is made up of fine pink macules & is found especially on the soles of the feet, wrists & forearms

P: oedema & necrosis

I: diagnosis usually clinical, serology; PCR

M: chloramphenicol, tetracycline or doxycycline

C: uraemia; DIC; pneumonia; otitis media; ileus; parotitis; meningoencephalitis; necrosis of digits

P: overall mortality 7–10% (25% at extremes of age); prevention – suitable clothing, repellant, avoidance, tick surveillance

D: causes a wide range of localised diseases of the respiratory & gastrointestinal tracts, musculoskeletal system & skin as well as septicaemia

A: *Staphylococcus* spp. are G +ve cocci; broadly divided into coagulase +ve & −ve; *S. aureus* is the most important coagulase ve type; *S. epidermidis* is the most common coagulase +ve; part of normal flora

A/R: broken skin, e.g. wounds, burns, skin disease; foreign material, e.g. IV catheter; damaged mucosal surfaces, e.g. 2° to viral infection

E: worldwide

H: ***S. aureus***

(1) localised infection
- (a) skin & appendages
  - (i) folliculitis – usually neck, axillae & buttocks → boils or carbuncles, often recurrent
  - (ii) impetigo – blistering skin lesions (often on face) with crusting exudates, occurs most often in children
  - (iii) paronychia
  - (iv) mastitis & breast abscess
- (b) ENT
  - (i) otitis externa – pain & itching
  - (ii) otitis media & sinusitis – much less common
- (c) wound infection
  - (i) most common cause of nosocomial wound infection – erythema & serous discharge → small abscesses (often around sutures) or → cellulitis, dehiscence, pain, systemic upset
  - (ii) IV devices – pyrexia or sepsis as early as 2/7 after insertion
  - (iii) prosthetic infection
- (d) pleuropulmonary
  - (i) aspiration pneumonia – generally complicates pre-existing lung disease
  - (ii) haematogenous spread – affects normal lungs; spread from skin, endocarditis, IV device; can be isolate bacteria from blood
  - (i) & (ii) severe disease → high fever & cyanosis
- (e) UTI – in instrumentation or catheterisation

(2) bacteraemia/septicaemia mostly 2° spread from local site but also from 1° abrasions → deep-seated infection involving joints, bones, lungs, heart valves

(3) haematogenous/metastatic infection
- (a) endocarditis – usually from a 1° community-acquired bacteraemia associated with asymptomatic congenital left valve abnormality, IVDA or IV devices; flu-like illness & GI upset → valvular insufficiency & emboli; 25% have meningism; also signs of emboli, chest signs; → emergency valve replacement
- (b) bone & joint infection – most commonly due to 1° bacteraemia or contiguous spread after trauma or surgery (especially prosthetic implants); vertebral column now most common site
- (c) pyomyositis – acute inflammation of skeletal muscle; ♂ » ♀; → pain, fever, induration, swelling; mainly in tropics & subtropics

(4) toxin-mediated
- (a) food poisoning (5% of outbreaks) – IP of hours → severe vomiting, nausea, abdominal cramps, diarrhoea

(b) SSSS – children > adults; sudden onset extensive erythema → bullous desquamation of large areas of skin

(c) TSS – associated with tampon use; → high fever, diarrhoea, confusion, erythroderma, hypertension, ARF; mortality 5%

Coagulase −ve

most infections are acquired in hospital

(1) IV devices – especially in neonates & compromised
(2) CSF shunts – low-grade fever, malaise, shunt malfunction; meningitic signs may be absent
(3) peritonitis associated with CAPD – abdominal pain, nausea, diarrhoea, fever
(4) endocarditis – native or prosthetic valve; nosocomial native valve disease generally from infected IV device; nosocomial prosthetic valve disease acquired in theatre or from an IV device; community-acquired native valve disease mimics *S. aureus* endocarditis
(5) UTI – *S. saprophyticus* in sexually active ♀ → cystitis but may cause UTI; nosocomial infection associated with surgery/catheterisation & usually *S. epidermidis*, may be multiply resistant
(6) other – bacteraemia in neonates & neutropaenics; commonest cause of post-operative endophthalmitis

E: see above

P: local or systemic inflammation; toxin production

I: depends on infection but consider – FBC – ↑ WBC ; ↑ ESR; ↑ CRP; stool/blood/dialysate – microscopy, culture & sensitivity; CSF – ↑ polymorphs; echocardiography; CXR, AXR

M: depends on infection but consider draining pus & removing foreign bodies; antibiotics depend on sensitivity but consider flucloxacillin if sensitive to methicillin (MRSA), vancomycin, rifampicin, doxycycline, linezolid, possibly in combination if resistant

C: abscess formation (*S. aureus*); endocarditis; bone or joint infection; bacteraemia

P: endocarditis may be life-threatening; correct management normally leads to excellent outcome

D: cause a wide range of diseases

A: *Streptococcus* spp. are G +ve cocci; there is no single ideal classification system for them; the most important are *S. pyogenes* (β-haemolytic group A), *S. agalactiae* (β-haemolytic group B), β-haemolytic groups C & G, viridans streptococci, *S. milleri* group, enterococci & *S. pneumoniae*

A/R: extremes of age, HIV/AIDS for *S. pneumoniae*

E: worldwide

H: ***S. pyogenes***

(a) pharyngitis/tonsillitis
IP 1–3/7 → sore throat, fever, headache, malaise, pain on swallowing, nausea, vomiting; cervical lymphadenopathy, red & oedematous pharynx, enlarged tonsils, spots & exudates; may → acute sinusitis, otitis media, abscess (Quinsy)

(b) SF & RhF
follows a streptococcal infection; mostly school age children; SF → diffuse symmetrical blanching erythematous rash; tongue strawberry → raspberry

(c) perianal infection
♂ > ♀; itching, pain on defecation, blood-stained stools; superficial well-demarcated rash spreading from anus

(d) vulvovaginitis
prepubescent ♀; serosanguinous discharge & erythema

(e) skin & soft tissues
- (i) pyoderma/impetigo & 2° infection in varicella; history of skin trauma; discrete purulent papules → vesicles → pustules
- (ii) erysipelas – acute inflammation of skin with lymph node involvement; sore throat, fever, rigors; local erythema → swells & spreads → vesicles & bullae & oedema, well-demarcated
- (iii) cellulitis – acute spreading infection of skin & subcutaneous tissues; history of mild trauma, burns, surgery, IVDA; pain, swelling, erythema, fever, rigors, malaise; lymphadenopathy
- (iv) necrotising fasciitis – infection of deep SC tissues & fascia; extensive rapidly spreading necrosis & gangrene; history of minor trauma; usually community-acquired; usually affects arm/leg; redness, swelling, fever, pain, purple discolouration & bullae; high mortality

(f) STSS
shock & multi-organ failure; usually associated with necrotising fasciitis or myositis; mostly community-acquired

(g) bacteraemia
community- & hospital-acquired; entry via skin; more cases of greater severity in those with underlying disease, e.g. malignancy, diabetes, immunocompromise

(h) others
puerperal sepsis

S. agalactiae (β-haemolytic group B)

(a) neonatal infection
bacteraemia +/– meningitis; also impetigo neonatorum, septic arthritis, osteomyelitis, peritonitis, pyelonephritis, facial cellulitis, conjunctivitis & endophthalmitis; early < 5 days after birth → mostly bacteraemia with high mortality; late 1–13/52 after birth → meningitis +/– bacteraemia, neurological sequelae common in survivors

(b) puerperal sepsis
1–2/7 postdelivery of abortion +/− retention of products; mostly endometritis with fever & uterine tenderness

(c) others
community-acquired disease in elderly skin & soft tissue infection occasional UTIs disease similar to *S. aureus*, e.g. osteomyelitis, endocarditis

β-haemolytic groups C & G
cause diseases like *S. pyogenes* except RhF

S. milleri group
abscess formation especially dental, intra-abdominal, liver, lung & brain; culture smells of caramel on blood agar

Viridans streptococci
biggest cause of infective endocarditis; *S. oralis* & *S. sanguis* are of oral or dental origin; *S. bovis* is a bowel commensal; may be resistant to penicillin; *S. bovis* associated with bowel cancer

Enterococci
E. avium, *E. casseliflavus*, *E. durans*, *E. faecalis* (most common) & *E. faecium*; nosocomial & colonisation; cause UTIs, wound infections, intra-abdominal infection, venflon-associated infection & dialysis-associated problems; resistant to many antibiotics so may need combination therapy

S. pneumoniae
worldwide causes 1 million childhood deaths p.a.; in temperate climates most disease is in winter; extremes of age most affected in industrialised countries; < 2 years of age & young adults in developing world; associated with diabetes, CHF, nephritic syndrome, HIV/AIDS, alcohol abuse, IVDA, congenital or other acquired immunocompromise

(a) pneumonia
fever, malaise, anorexia, weakness, headache, myalgia, chest pain, cough; fever, toxaemia, signs of lobar consolidation, cyanosis, confusion, jaundice; overall fatality 5%; complications include pleural & pericardial effusion & empyema

(b) otitis media
fever, ear pain +/− deafness/tinnitus, irritability; red bulging tympanic membrane +/− perforation; complications include chronic discharging ear, mastoiditis, meningitis, cerebral abscess

(c) meningitis
fever, nausea, backache, neckache, photophobia, failure to feed (infants); fever, toxic, stiff neck, +ve Kernig's, impaired consciousness, cranial nerve palsies (III & VI); 30–50% mortality within 1–2/7; 50% of survivors have neurological sequelae

E: see above

P: pus forming, toxin production, superantigen

I: FBC – ↑ WBC; ↑ CRP; may have ↑ LFT; swab/blood/sputum – microscopy, culture & sensitivity; CSF – turbid, ↑ WBC, ↑ protein, ↓ glucose

M: antibiotic depends on site & sensitivity – penicillin or gentamicin or ceftriaxone or other agents; may need ITU; surgical debridement; O_2 therapy; fluid balance, steroids for meningitis

C: see above

P: see above

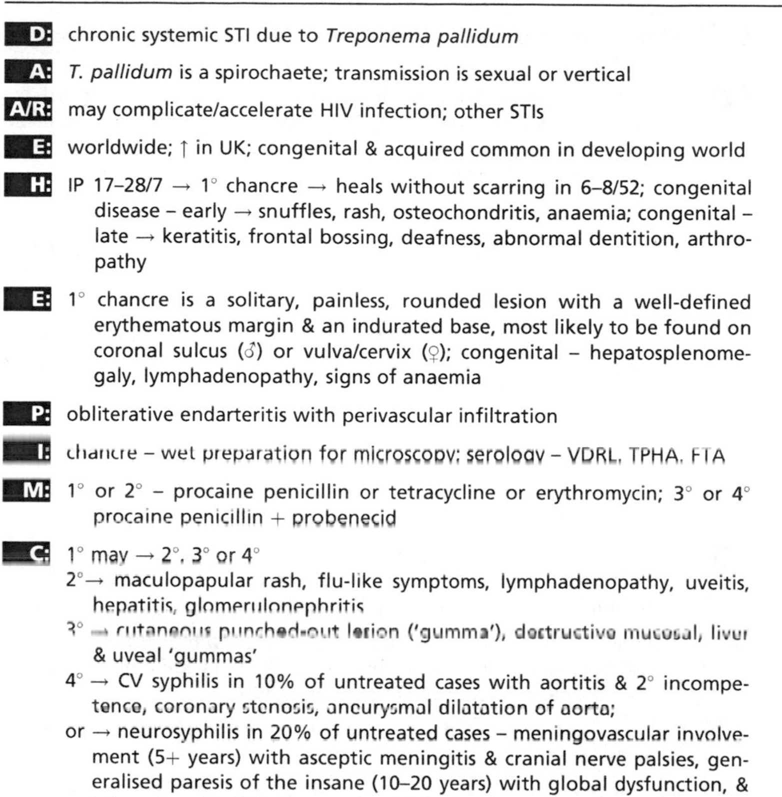

D: chronic systemic STI due to *Treponema pallidum*

A: *T. pallidum* is a spirochaete; transmission is sexual or vertical

A/R: may complicate/accelerate HIV infection; other STIs

E: worldwide; ↑ in UK; congenital & acquired common in developing world

H: IP 17–28/7 → 1° chancre → heals without scarring in 6–8/52; congenital disease – early → snuffles, rash, osteochondritis, anaemia; congenital – late → keratitis, frontal bossing, deafness, abnormal dentition, arthropathy

E: 1° chancre is a solitary, painless, rounded lesion with a well-defined erythematous margin & an indurated base, most likely to be found on coronal sulcus (♂) or vulva/cervix (♀); congenital – hepatosplenomegaly, lymphadenopathy, signs of anaemia

P: obliterative endarteritis with perivascular infiltration

I: chancre – wet preparation for microscopy; serology – VDRL, TPHA, FTA

M: 1° or 2° – procaine penicillin or tetracycline or erythromycin; 3° or 4° procaine penicillin + probenecid

C: 1° may → 2°, 3° or 4°

2°→ maculopapular rash, flu-like symptoms, lymphadenopathy, uveitis, hepatitis, glomerulonephritis

3° → cutaneous punched-out lesion ('gumma'), destructive mucosal, liver & uveal 'gummas'

4° → CV syphilis in 10% of untreated cases with aortitis & 2° incompetence, coronary stenosis, aneurysmal dilatation of aorta;

or → neurosyphilis in 20% of untreated cases – meningovascular involvement (5+ years) with asceptic meningitis & cranial nerve palsies, generalised paresis of the insane (10–20 years) with global dysfunction, & tabes dorsalis (15–35 years) with pain, loss of sensory modalities, hypotonia, ataxia, Charcot's joints, loss of tendon reflexes & bladder disturbances

P: good if treated; prevention – contact tracing

Tapeworms

D: infection with tapeworms (cestodes) produces a range of diseases and symptoms

A: Cestodes are hermaphrodite flatworms; most common are *Diphyllobothrium latum* (fish tapeworm), *Taenia saginata* (beef) & *Taenia solium* (pork, cysticercosis); humans act as definitive or intermediate hosts by ingesting worms or ova respectively

A/R: eating raw or undercooked fish/beef/pork

E: definitive disease – Scandinavia & Far East (fish), tropics (beef) & worldwide (pork); intermediate disease – *T. solium*; hydatid disease *E. echinococcus granulosus* (dog)

H: definitive disease may be asymptomatic or cause vague abdominal pain; intermediate disease – ingestion of cysts → fever, muscle ache → years of latency → cysts die, may cause fits if in CNS

E: may see motile segments of *T. saginata* in anus; hydatid disease is characterised by a palpable, slow-growing liver 'tumour'

P: fish – worms compete for dietary B12; cysticercosis – inflammation & fibrosis, calcification when dead

I: definitive – FBC – ↑ eosinophils; FBC – ↓ Hb due to ↓ B12 in fish tapeworm; stool – microscopy for ova or segments; intermediate – FBC – ↑ eosinophils; CXR, CT, MRI or USS to look for lesions

M: definitive – niclosamide (+ saline purge in *T. solium* to prevent cysticercosis); cysticercosis – anticonvulsants +/– praziquantel; hydatid – albendazole in 3 cycles of 28/7, cyst injection

C: megaloblastic anaemia (fish); recurrent fits (cysticercosis); cysts rupture (hydatid)

P: important cause of anaemia; may not respond well to treatment; re-infection common; prevention – treatment of raw sewage, meat inspection, adequate cooking/freezing of meat

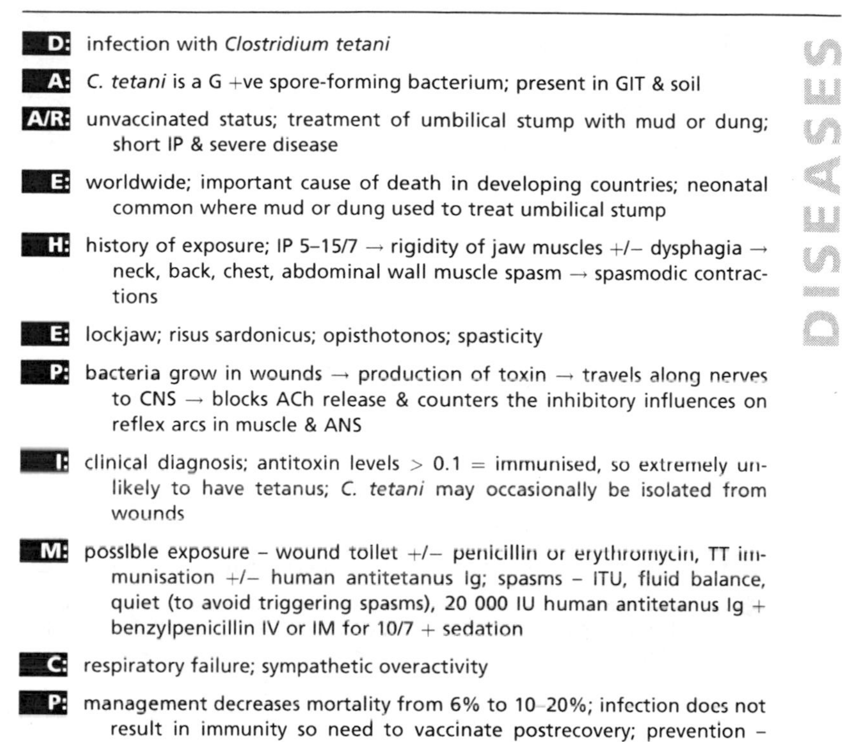

Tetanus

D: infection with *Clostridium tetani*

A: *C. tetani* is a G +ve spore-forming bacterium; present in GIT & soil

A/R: unvaccinated status; treatment of umbilical stump with mud or dung; short IP & severe disease

E: worldwide; important cause of death in developing countries; neonatal common where mud or dung used to treat umbilical stump

H: history of exposure; IP 5–15/7 → rigidity of jaw muscles +/– dysphagia → neck, back, chest, abdominal wall muscle spasm → spasmodic contractions

E: lockjaw; risus sardonicus; opisthotonos; spasticity

P: bacteria grow in wounds → production of toxin → travels along nerves to CNS → blocks ACh release & counters the inhibitory influences on reflex arcs in muscle & ANS

I: clinical diagnosis; antitoxin levels > 0.1 = immunised, so extremely unlikely to have tetanus; *C. tetani* may occasionally be isolated from wounds

M: possible exposure – wound toilet +/– penicillin or erythromycin, TT immunisation +/– human antitetanus Ig; spasms – ITU, fluid balance, quiet (to avoid triggering spasms), 20 000 IU human antitetanus Ig + benzylpenicillin IV or IM for 10/7 + sedation

C: respiratory failure; sympathetic overactivity

P: management decreases mortality from 6% to 10–20%; infection does not result in immunity so need to vaccinate postrecovery; prevention – active vaccination

Tick-borne encephalitis

D: 3 major diseases are RSSE, ETBE & CTF

A: all are arbovirus infections; transmitted by *Ixodes* from small wild animals; also from contaminated goat's milk

A/R: mortality higher in children; travel to affected area; forest work

E: RSSE & ETBE seasonally epidemic in former USSR, E. Europe, Scandinavia; CTF in Rocky Mountain area

H: often asymptomatic; RSSE & ETBE IP 8–14/7 → severe headache, fever, nausea → subsides in 7/7 or may become biphasic → meningoencephalitis; CTF biphasic

E: 2nd phase – meningoencephalitis signs (drowsiness, irritability, etc.), ascending paralysis, respiratory distress; CTF – may have maculopapular or petechial rash

P: multiples in liver → brain via blood → severe neuronal damage in cervical cord, medulla, midbrain & pons

I: CSF – ↑ WBC, ↑ protein; blood – isolate virus; serology

M: supportive; hyperimmune serum may be helpful in 1st week

C: residual paralysis of upper extremities/shoulder girdle

P: severe RSSE & ETBE have up to 30% & 3% mortality respectively; sequelae uncommon in CTF; prevention – vaccine, tick repellant & suitable clothing

D: infection with *Toxoplasma gondii* causing a range of disease including hepatitis, pneumonitis & CNS infection

A: *T. gondii* is an obligate intracellular pathogen; transmission is with oocysts from cat faeces or cysts from undercooked meat

A/R: HIV/AIDS; other immunocompromised; pregnancy

E: worldwide

H: competent hosts → asymptomatic or rarely a mononucleosis-like illness; compromised host → fever, cough, dyspnoea (pneumonitis) or speech abnormality, altered higher function (CNS)

E: competent – lymphadenopathy; compromised – chest signs or focal neurological signs including cranial nerve lesion & coma

P: stellate abscess formation; parasites remain viable until death of host

I: blood/body fluids/lymph node – microscopy; serology – IgM or IgA presence of 4 × ↑ or ↓ in IgG, PCR

M: pyrimethamine + sulfadiazine + Ca^{2+} folinate; maintenance doses for life

C: compromised host – lung, heart, chorioretinal involvement; pregnant – serious congenital abnormalities

P: CNS disease has 10% mortality rate & 10% of survivors have serious neurological complications; prevention – improved food hygiene/preparation, avoid cats in pregnancy & immunocompromise

Treponematosis

D: infection with non-venereal type *Treponema* spp. causing yaws, pinta & bejel

A: *Treponema* spp. are spirochaetes; *T. pallidum* causes bejel (endemic syphilis, firjal) which is transmitted by direct & indirect contact with an infected person; *T. carateum* causes pinta (azul, carate, mal de pinto) transmitted by lesion-to-skin contact; *T. pertenue* causes yaws (buba, framinosa, pian) transmitted by lesion-to-broken skin contact

A/R: age 2–15 years; overcrowding; poverty; poor sanitation

E: bejel – W. Africa, nomadic Arabians, C. Australian aborigines; pinta – Mexico, northern S. America; yaws – Africa, C. & S. America, Indonesia, Papua New Guinea, parts of India & Thailand

H: bejel – 1° lesion rare; later → mucosal patches, sore throat, hoarseness → healing, may also have bone pain, papilloma, angular stomatitis, rash; pinta – 1° lesion @ site of entry → 2° lesions several months later, itchy plaques 'pintids'; yaws – IP 21/7 → early – mother yaw @ site of entry, itchy, heals; may → 1° in 3–6/12 or 2° up to 24/12 → crop of lesions, may → spontaneous cure or latency or late disease

E: bejel: lymphadenopathy & shallow, painless mucosal ulcers; pinta: 1° – erythematous papule with satellites, 2° – 'pintids' of various colours/sizes; yaws: mother yaw – on face, legs, arms or neck = round/oval papule, 2° – multiple papilloma on any part of body, late – lesions similar to venereal syphilis, also hyperkeratosis of palms/soles, bursitis, disfiguring lesions of nasopharynx

P: inflammation

I: exudates – microscopy; serology as venereal syphilis – VDRL

M: consider penicillin G or erythromycin if allergic

C: bejel – late lesions like yaws; pinta – hyperchromic/atrophic skin; yaws – bone involvement

P: bejel – relapse rare; yaws – may relapse for up to 10 years

Trichomoniasis

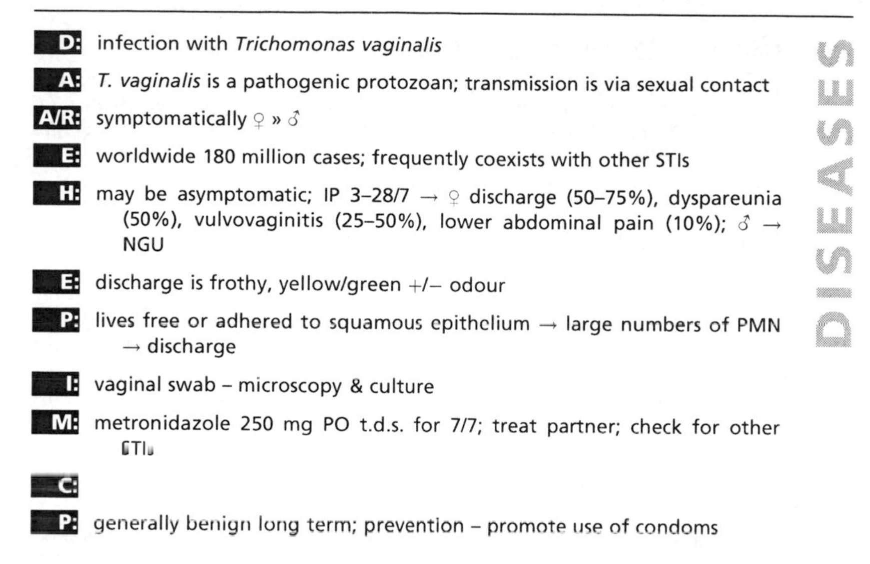

D: infection with *Trichomonas vaginalis*

A: *T. vaginalis* is a pathogenic protozoan; transmission is via sexual contact

A/R: symptomatically ♀ » ♂

E: worldwide 180 million cases; frequently coexists with other STIs

H: may be asymptomatic; IP 3–28/7 → ♀ discharge (50–75%), dyspareunia (50%), vulvovaginitis (25–50%), lower abdominal pain (10%); ♂ → NGU

E: discharge is frothy, yellow/green +/– odour

P: lives free or adhered to squamous epithelium → large numbers of PMN → discharge

I: vaginal swab – microscopy & culture

M: metronidazole 250 mg PO t.d.s. for 7/7; treat partner; check for other STIs

C:

P: generally benign long term; prevention – promote use of condoms

Trypanosomiasis

D: infection with *Trypanosoma* spp.

A: *Trypanosoma* spp. are flagellate protozoa; *T. brucei* causes sleeping sickness (African) transmitted by tsetse fly (*Glossina* spp.); *T. cruzi* causes Chagas' disease (American) transmitted by *Triatoma* & *Rhodnius prolixus*

A/R: presence of vector

E: African – endemic disease in E. & S. Africa (*T. b. rhodesiense*), epidemic disease in W. & C. Africa (*T. b. gambiense*); American – N.W.

H: African – IP hours–weeks → localised rash, pain, heat, periodic fever, headache, joint pain, myalgia, weight loss, diarrhoea, pruritis, hyper- or paraesthesia, insomnia or somnolence, ataxia, paralysis, altered speech

American – acute → fever & peripheral oedema; chronic → cardiac & GIT involvement

E: African – lymphadenopathy, oedema, signs of CHF, endocrine disturbance, altered reflexes, fits, mental disorder, ataxia/dyskinesia, paralysis, altered speech

American – acute → fever, oedema, lymphadenopathy, hepatosplenomegaly; chronic → signs of cardiomyopathy or megaoesophagus (5%) or megacolon (« 5%)

P: African – escapes immune system by varying surface coat, spreads via blood & lymphatics, causes haemorrhage & tissue oedema

American – all tissues nested but predilection for heart muscle causing oedema

I: African – FBC – ↓ Hb; stained thick blood film after centrifugation; ECG – abnormalities; serological testing to screen populations

American – FBC – ↑ WBC, ↓ Hb; ↑ LFT; ↑ CK; (acute) stained thick blood film after centrifugation; ECG –abnormalities; (chronic) serology at reference laboratory

M: African – requires specialist knowledge as chemotherapeutics complicated & potentially very toxic

American – nifurtimox or benzimidazole; symptomatic management of cardiac problems & megasyndromes

C: African – severe CNS involvement; toxicity & side-effects of treatment;

American – CNS damage; exocrine abnormalities; volvulus; aspiration pneumonia

P: African – mortality high without treatment; 90% cure if effectively managed; no good evidence for immunity; relapse & sequelae unusual; prevention – control vector with flytraps, screens etc.

American – cardiac failure has 50% 2-year survival; prevention – control vector

D: infection with *Mycobacterium tuberculosis* causes 1° disease & complications of reactivation

A: *M. tuberculosis* is an acid-fast bacillus; transmission is via inhalation of droplets

A/R: 75% of 1° infections in 15–50-year age group; HIV/AIDS; other immunocompromised; alcohol; smoking; genetics; malnutrition; poverty; travel; war; poor drug compliance; unvaccinated

E: worldwide; 30% of total population infected; 3 million deaths p.a.

H: 1° symptomatic or fever, malaise, cough → resolution in competent host or disseminated in compromised; post 1° (10% of competent) → weight loss, anorexia, fever, malaise, night sweats, more specific symptoms depending on location:
pulmonary – productive cough, haemoptysis, chest pain, breathlessness
miliary (infants & compromised) – gradual onset fever, malaise, weight loss
meningitis – headache, vomiting, ↓ consciousness, seizures
adenitis – lymphadenopathy
skeletal – deformity +/– paraplegia
GI – diarrhoea, abdominal pain, distension
pericardial (especially in HIV/AIDS) – fever, chest pain genitourinary – symptoms of UTI, abdominal pain, infertility

E: 1° may have erythema nodosum; post 1° depends on location:
pulmonary – ill, wasting, fever, tachycardia, fine creps, bronchial breathing, wheeze, signs of pleural effusion
miliary – hepatomegaly, splenomegaly, neck stiffness
meningitis – III, IV, VI & VIII nerve palsies
adenitis – lymphadenopathy most usually cervical (scrofula)
skeletal: usually affects spine (Potts)
GI – ascites, fistulae
pericardial – pericardial rub
genitourinary – epididymal swelling, endometritis

P: granulomatous response

I: 1° – tuberculin test conversion (Heaf or Mantoux; little use in compromised or vaccinated); CXR may show consolidation +/– lymphadenopathy; post 1° depends on location:
pulmonary – 3× sputum – Z–N for AFB & culture (slow); CXR
miliary – 3× sputum – Z–N for AFB (but often –ve); tuberculin (but often –ve); CXR; biopsy may yield AFB
meningitis – CSF – ↑ WBC, ↑ protein, ↓ glucose, Z–N for AFB, culture & sensitivity
adenitis – biopsy – Z–N for AFB & culture & sensitivity
skeletal – X ray; biopsy – Z–N for AFB & culture & sensitivity
pericardial – echo; pericardial tap – Z–N for AFB & culture & sensitivity
genitourinary – urinalysis – ↑ WBC, haematuria, but –ve for nitrites, culture & sensitivity; AXR

M: treat for at least 6/12; consider DOTS; many regimens, e.g. isoniazid + rifampicin + ethambutol for 2/12 then isoniazid + rifampicin for 4/12; 2nd line drugs include ethionamide, amikacin, streptomycin, quinolones, clarithromycin, capreomycin, clofazimine, cycloserine, rifabutin; surgery may be useful, steroids for meningitis and pericarditis

DISEASES

C: 1° – meningitis or miliary; all – local infiltration & destruction, MDRTB; pulmonary – pleural effusion, empyema, pleurisy, pneumothorax, cor pulmonale, bronchiectasis, 2° infection of cavities, dissemination

P: good with treatment; miliary poorer prognosis; prevention – vaccine gives 0–80% protection, socio-economic development, treatment of underlying immunocompromise, education, chemoprophylaxis

D: infection with *Francisella tularensis* types A & B; may be used as a biological weapon

A: *F. tularensis* is a G −ve bacterium; causes disease in humans & rodents; transmission is by vector, contaminated water, infected meat, aerosol & dust

A/R: occupational exposure – laboratory work, skin trappers, agricultural workers, soldiers

E: N. America, Europe, former USSR, Japan; severity of type A (rabbit-borne) $>$ type B (rat-borne)

H: may be asymptomatic; IP 1–10/7 → cutaneous (60%) = papule @ site of infection (bite or abrasion) → pustule → necrosis → ulcer +/− fever, prostration; or pneumonic/typhoidal (16%) → sudden onset severe headache, vomiting, chills, fever, dyspnoea, pleuritic chest pain, myalgia, sweats, prostration, loss of weight; or ophthalmic (1%) → unilateral itching, lacrimation, pain, photophobia

L: cutaneous – lymphadenopathy, papule/pustule/punched-out ulcer; pneumonic – temperature $>$ 40°C, generalised weakness, petechial/papular/roseolar/pustular rash, tender splenomegaly (30%); ophthalmic – head/neck lymphadenopathy, swollen eyelids, red conjunctivae with small nodes/grey exudate/ulcers

P: haemorrhagic oedema & necrosis

I: FBC – ↔ or slightly ↑ WBC; aspirate/biopsy – culture

M: streptomycin or gentamicin + chloramphenicol if meningitis; ? role for ciprofloxacin

C: pulmonary – meningitis, pericarditis, pulmonary abscess; ophthalmic – permanent visual impairment; Jarisch–Herxheimer reaction

P: if untreated up to 60% mortality; rarely fatal if treated; prevention – vaccine decreases severity of disease, avoid infection

Typhoid & paratyphoid

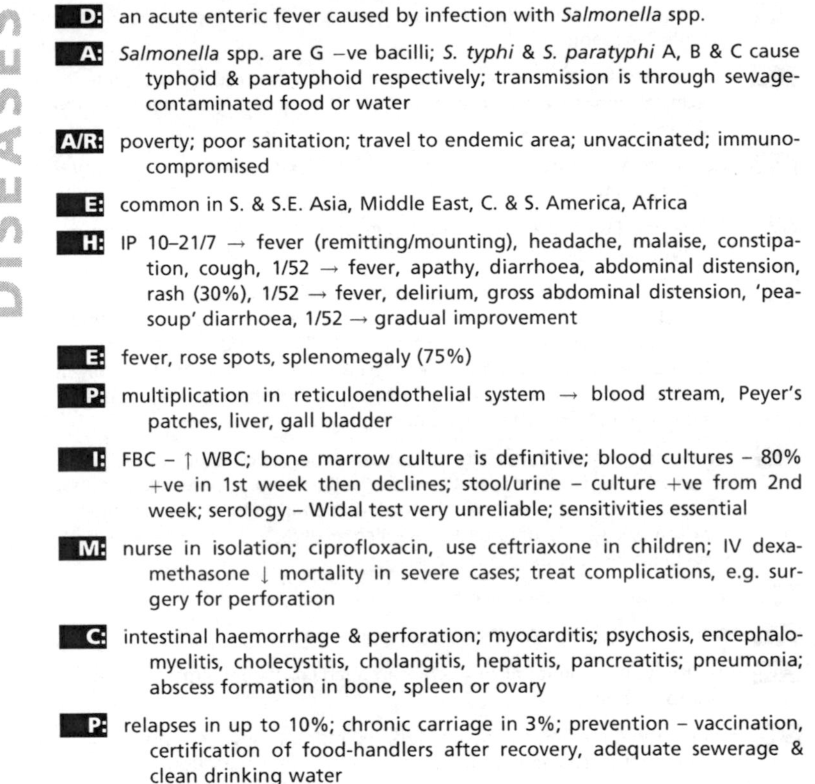

D: an acute enteric fever caused by infection with *Salmonella* spp.

A: *Salmonella* spp. are G −ve bacilli; *S. typhi* & *S. paratyphi* A, B & C cause typhoid & paratyphoid respectively; transmission is through sewage-contaminated food or water

A/R: poverty; poor sanitation; travel to endemic area; unvaccinated; immunocompromised

E: common in S. & S.E. Asia, Middle East, C. & S. America, Africa

H: IP 10–21/7 → fever (remitting/mounting), headache, malaise, constipation, cough, 1/52 → fever, apathy, diarrhoea, abdominal distension, rash (30%), 1/52 → fever, delirium, gross abdominal distension, 'pea-soup' diarrhoea, 1/52 → gradual improvement

E: fever, rose spots, splenomegaly (75%)

P: multiplication in reticuloendothelial system → blood stream, Peyer's patches, liver, gall bladder

I: FBC – ↑ WBC; bone marrow culture is definitive; blood cultures – 80% +ve in 1st week then declines; stool/urine – culture +ve from 2nd week; serology – Widal test very unreliable; sensitivities essential

M: nurse in isolation; ciprofloxacin, use ceftriaxone in children; IV dexamethasone ↓ mortality in severe cases; treat complications, e.g. surgery for perforation

C: intestinal haemorrhage & perforation; myocarditis; psychosis, encephalomyelitis, cholecystitis, cholangitis, hepatitis, pancreatitis; pneumonia; abscess formation in bone, spleen or ovary

P: relapses in up to 10%; chronic carriage in 3%; prevention – vaccination, certification of food-handlers after recovery, adequate sewerage & clean drinking water

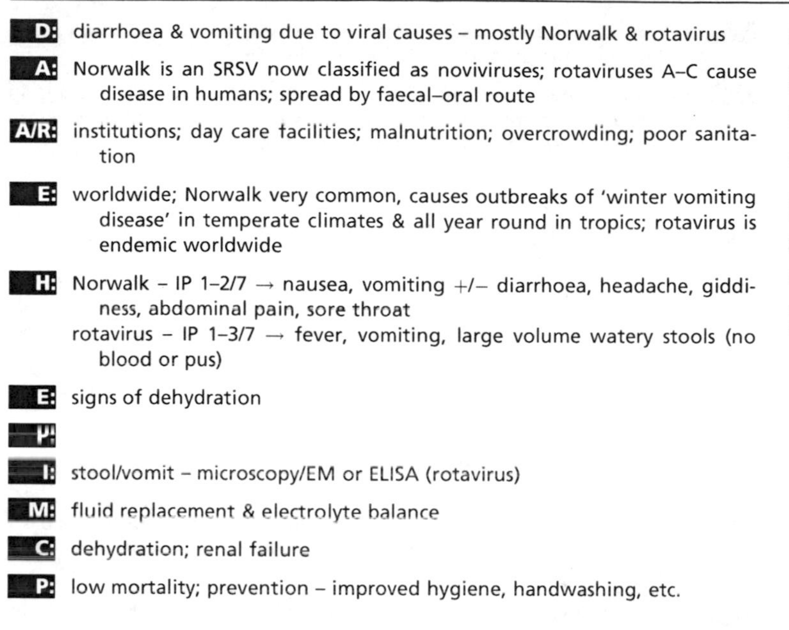

D: diarrhoea & vomiting due to viral causes – mostly Norwalk & rotavirus

A: Norwalk is an SRSV now classified as noviviruses; rotaviruses A–C cause disease in humans; spread by faecal–oral route

A/R: institutions; day care facilities; malnutrition; overcrowding; poor sanitation

E: worldwide; Norwalk very common, causes outbreaks of 'winter vomiting disease' in temperate climates & all year round in tropics; rotavirus is endemic worldwide

H: Norwalk – IP 1–2/7 → nausea, vomiting +/– diarrhoea, headache, giddiness, abdominal pain, sore throat
rotavirus – IP 1–3/7 → fever, vomiting, large volume watery stools (no blood or pus)

E: signs of dehydration

P:

I: stool/vomit – microscopy/EM or ELISA (rotavirus)

M: fluid replacement & electrolyte balance

C: dehydration; renal failure

P: low mortality; prevention – improved hygiene, handwashing, etc.

Viral haemorrhagic fevers

D: VHF are caused by a number of different viruses including Ebola, Lassa & Marburg that produce mild to severe disease

A: category 4 viral pathogens; Ebola & Marburg are filoviruses; Lassa is an arenavirus; transmission is by contact with infected people/tissues/secretions or rat urine (Lassa) or monkeys (Marburg)

A/R: travel to endemic area; exposure to infected people/tissues, e.g. health care professionals

E: Ebola & Marburg – Africa; Lassa – W. Africa; both cause outbreaks

H: history of exposure or travel to region; IP $<$ 3/52 → fever, pharyngitis, conjunctivitis, vomiting, diarrhoea, abdominal pain; Ebola after 5/7 most cases → bleeding +/– psychosis, hemiplegia; Lassa after 1/52 minority → oedema, bleeding from mucous membranes

E: signs of bleeding, dehydration, shock; Ebola & Marburg – rash, neurological manifestations; Lassa – hypotension, bradycardia, oedema

P: ↑ vascular permeability & immune complex deposition → haemorrhage & necrosis

I: all – take extreme care with specimens; rule out other causes; blood – ↓ WBC, ↑ AST, ↓ protein, urinalysis – albuminuria; throat swabs/MSU/blood for EM; serology

M: barrier nursing; fluid & electrolyte balance; Lassa – role for ribavirin in 1st week; ? role for convalescence serum

C: widespread organ damage – liver, lungs, heart, CNS; shock

P: Ebola mortality 50–80%; Lassa haemorrhagic disease has up to 30% mortality; Marburg may relapse; prevention – rodent control (Lassa), careful management of cases to prevent additional cases

Visceral larva migrans

D: caused by *Toxocara canis*

A: *T. canis* is a roundworm infection of dogs, arrested in larval stage in humans; transmission is from dog faeces, contaminated soil & puppies

A/R: young children; pica (soil eating); heavy infections are more severe

E: USA, Europe, Caribbean, Mexico, Hawaii, Philippines, Australia, S. Africa

H: contact with puppies; many infections are asymptomatic; or → malaise, fever, asthma, rash

E: characteristic rash; hepatomegaly; pulmonary signs; cardiac dysfunction; neurological lesions

P: arrest at larval stage causes formation of granulomatous lesions in liver

I: FBC – ↑ WBC (eosinophilia); ↓ albumin : globulin ratio; ELISA; CXR – mottling

M: albendazole or mebendazole are controversial; consider steroids

C: granulomatous lesions may also form in lungs, heart, kidneys, muscle, brain & eye; massive larval ingestion may rarely be fatal

P: most recover naturally after 2 years; no relapses & 2° infection unlikely; prevention – treat dogs for infection, avoid children's contact with puppies, ban dogs from playgrounds/sandpits

Whooping cough

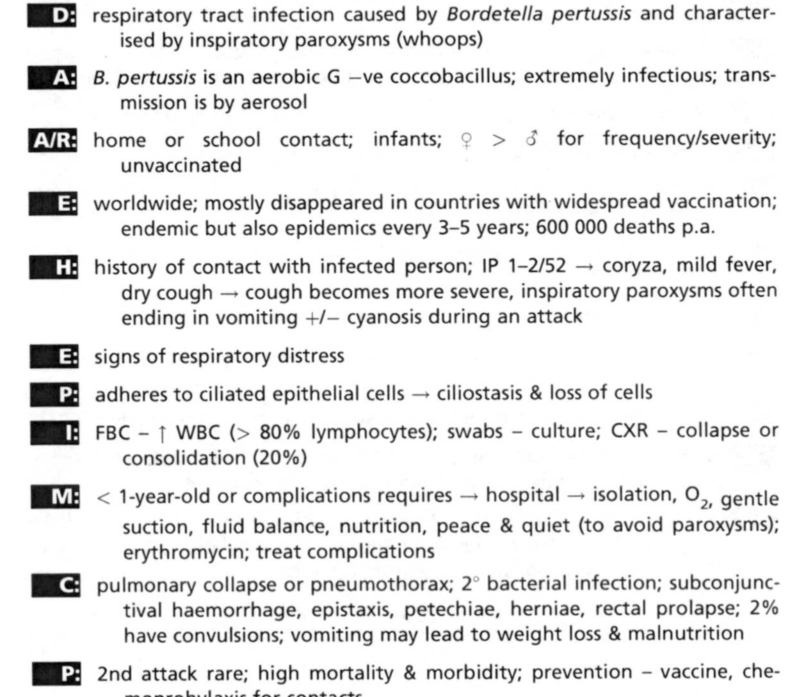

D: respiratory tract infection caused by *Bordetella pertussis* and characterised by inspiratory paroxysms (whoops)

A: *B. pertussis* is an aerobic G −ve coccobacillus; extremely infectious; transmission is by aerosol

A/R: home or school contact; infants; ♀ > ♂ for frequency/severity; unvaccinated

E: worldwide; mostly disappeared in countries with widespread vaccination; endemic but also epidemics every 3–5 years; 600 000 deaths p.a.

H: history of contact with infected person; IP 1–2/52 → coryza, mild fever, dry cough → cough becomes more severe, inspiratory paroxysms often ending in vomiting +/− cyanosis during an attack

E: signs of respiratory distress

P: adheres to ciliated epithelial cells → ciliostasis & loss of cells

I: FBC – ↑ WBC (> 80% lymphocytes); swabs – culture; CXR – collapse or consolidation (20%)

M: < 1-year-old or complications requires → hospital → isolation, O_2, gentle suction, fluid balance, nutrition, peace & quiet (to avoid paroxysms); erythromycin; treat complications

C: pulmonary collapse or pneumothorax; 2° bacterial infection; subconjunctival haemorrhage, epistaxis, petechiae, herniae, rectal prolapse; 2% have convulsions; vomiting may lead to weight loss & malnutrition

P: 2nd attack rare; high mortality & morbidity; prevention – vaccine, chemoprohylaxis for contacts

D: zoonotic VHF of monkeys & humans caused by YF virus; 'jungle' & 'urban' disease

A: YF virus is a flavivirus; transmission is via mosquitoes, mainly *Aëdes aegypti*

A/R: travel to endemic area; unvaccinated; forest clearing ('jungle'); dense population ('urban')

E: endemic in Africa & S. America, 200 000 cases/year

H: commonly inapparent infection; IP 3–6/7 → acute biphasic fever, flu-like symptoms, conjunctivitis → remission for up to 7/7, may then → vomiting, abdominal pain, jaundice, haemorrhage

E: fever; jaundice, hypotension, haemorrhage

P: midzone hepatic necrosis & ATN

I: FBC – ↓ WBC, ↓ platelets; serology – ELISA, IgM or 4× ↑ IgG between acute & convalescence serum

M: supportive treatment

C: DIC; shock; convulsions, coma; hepatitis; myocarditis

P: if jaundiced, mortality 20–60%; prevention – vaccine

Yersiniosis

D: infection with non-*pestis Yersinia* spp.

A: *Yersinia* spp. are G −ve bacilli; *Y. enterocolitica* causes diarrhoea; *Y. pseudotuberculosis* causes adenitis; transmission is via ingestion of contaminated pork, water or milk

A/R: < 5-year-olds; immunocompromised; iron overload; HLA-B27 (*Y. enterocolitica*)

E: both worldwide; *Y. enterocolitica* is more common in temperate zones

H: *Y. enterocolitica*→ nausea, abdominal pain, nausea, vomiting
Y. pseudotuberculosis→ abdominal pain

E: *Y. enterocolitica*→ abdominal tenderness
Y. pseudotuberculosis→ pseudoappendicitis – fever, RIF pain

P: *Y. pseudotuberculosis* causes mesenteric adenitis

I: stool – microscopy & culture; serology

M: *Y. enterocolitica* needs no treatment if mild, but consider co-trimoxazole or ciprofloxacin in severe cases
Y. pseudotuberculosis may end up being managed as appendicitis

C: *Y. enterocolitica* – septicaemia; reactive arthritis; erythema nodosum
Y. pseudotuberculosis – GIT ulceration, perforation, intussusception, toxic megacolon, cholangitis, mesenteric vein thrombosis

P: *Y. enterocolitica*+ cirrhosis/alcohol abuse/immunocompromise/iron overload has up to 50% mortality; infection results in immunity

APPENDICES

Vaccines	
Anthrax	
Vaccine type	Cell-free filtrate
Indications for use	At-risk groups including laboratory and hide workers, and military personnel
Schedule • number of doses • timing • booster	 6 SC 0, day 2, day 4, 6/12, 12/12 & 18/12 annually
Contraindications & Precautions	Pregnancy is a contraindication
Reactions/Side-effects	Mild local reaction
Additional notes	
Diphtheria	
Vaccine type	Formaldehyde-inactivated toxin (absorbed onto aluminium sulphate)
Indications for use	EPI Routine childhood vaccination
Schedule • number of doses • timing • booster	 3, 4 or 5 IM 3 = during 1st year of life 4 = 3 during 1st year of life + booster around 2nd/3rd year 5 = 3 during 1st year of life + booster at 2 years + booster on entering school 10 years
Contraindications & Precautions	Hypersensitivity reaction to previous dose is a contraindication
Reactions/Side-effects	Mild local or systemic reaction common
Additional notes	Can be D, DT, DTP or DTP-HepB-Hib Considerable variation between national vaccination schedules

(*continued*)

Vaccines *continued*

Haemophilus influenzae b

Vaccine type	Conjugate
Indications for use	Routine childhood vaccination
Schedule • number of doses • timing • booster	 2 or 3 (depending on manufacturer) IM 6/52, 10/52 & 14/52 of age none
Contraindications & Precautions	Hypersensitivity reaction to previous dose is a contraindication
Reactions/Side-effects	Mild local reaction
Additional notes	

Hepatitis A

Vaccine type	Formaldehyde-inactivated virus Live attenuated under development
Indications for use	Travel to endemic areas
Schedule • number of doses • timing • booster	 2 IM 12/12 apart 10 years
Contraindications & Precautions	Hypersensitivity reaction to previous dose is a contraindication
Reactions/Side-effects	Mild local and systemic reaction
Additional notes	Can be combined with hepatitis B

Hepatitis B

Vaccine type	Inactivated HepB surface antigen (absorbed onto aluminium salts) Also recombinant DNA and plasma-derived HepB surface antigens
Indications for use	EPI Inclusion in national immunisation programmes ideal Exposed and high-risk populations, e.g. health care workers

Schedule • number of doses • timing • booster	 3 IM 0, 1/12 & 6/12 10 years
Contraindications & Precautions	Hypersensitivity reaction to previous dose is a contraindication
Reactions/Side-effects	Local reaction
Additional notes	Avoid injecting into buttock as it decreases efficacy Can also be DTP-HepB or DTP-HepB-Hib

Influenza A & B

Vaccine type	Inactivated virus (grown in eggs)
Indications for use	At-risk populations: elderly, chronic respiratory disease and other chronic disease; diabetes and other endocrine disorders; immunosuppression
Schedule • number of doses • timing • booster	 1 SC or IM annually, start of flu season
Contraindications & Precautions	Egg allergy is a contraindication
Reactions/Side-effects	Mild local and systemic reaction
Additional notes	WHO recommends strains for inclusion each year

Japanese B encephalitis

Vaccine type	Formalin-inactivated (derived from mouse brain)
Indications for use	Travel to endemic areas
Schedule • number of doses • timing • booster	 Standard 3 or reduced 2 SC 0, day 7 & day 28 1 year then 3-yearly
Contraindications & Precautions	Hypersensitivity reaction to previous dose is a contraindication

(*continued*)

Vaccines *continued*

Japanese B encephalitis

Reactions/Side-effects	Mild local or systemic reaction Occasional severe reaction – generalised urticaria, hypotension and collapse
Additional notes	Encourage avoidance of mosquito bites

Measles

Vaccine type	Live attenuated virus
Indications for use	EPI Routine infant vaccination
Schedule • number of doses • timing • booster	 1 IM or SC with opportunity for 2nd 9–11/12 of age in highly endemic areas 12–15/12 of age in controlled areas 6–9/12 of age extra dose for very high risk, e.g. refugee camps 3–5 years of age
Contraindications & Precautions	Hypersensitivity reaction to previous dose is a contraindication Pregnancy and congenital/acquired immune disorders are contraindications HIV is *not* a contraindication
Reactions/Side-effects	Malaise, fever & rash 5–12/7 later Idiopathic thrombocytopaenic purpura Rarely encephalopathy and anaphylaxis
Additional notes	Can be given as MMR

Meningococcus

Vaccine type	Purified capsular polysaccharide
Indications for use	Emergency, e.g. outbreak High risk, e.g. travellers, army Routine childhood vaccination
Schedule • number of doses • timing • booster	 1 IM 3–5-yearly
Contraindications & Precautions	Hypersensitivity reaction to previous dose is a contraindication

Reactions/Side-effects	Mild local reaction, mild fever
Additional notes	Can be A or C, A & C, or A & C & W135 & Y, A

MenC

Vaccine type	Purified capsular polysaccharide conjugated to a protein
Indications for use	Emergency, e.g. outbreak High risk, e.g. travellers, army Routine childhood vaccination
Schedule • number of doses • timing • booster	 1 in older children, 3 in infants IM none
Contraindications & Precautions	Hypersensitivity reaction to previous dose is a contraindication
Reactions/Side-effects	Mild local reaction
Additional notes	Can be combined with other vaccines Not effective in children < 2 years of age

Mumps

Vaccine type	Live attenuated virus
Indications for use	Routine infant/childhood vaccination
Schedule • number of doses • timing • booster	 1 SC 9–12/12 of age in highly endemic areas 12–15/12 of age in controlled areas true booster not required
Contraindications & Precautions	Hypersensitivity reaction to previous dose or allergy to vaccine components are contraindications Advanced immune deficiency or suppression is a contraindication Avoid in pregnancy
Reactions/Side-effects	Parotitis and low-grade fever

(*continued*)

APPENDICES

Vaccines *continued*

Mumps	
	Aseptic meningitis occurs at widely different frequencies
Additional notes	Can be given as M, MM or MMR

Pertussis	
Vaccine type	Whole cell killed Acellular
Indications for use	EPI Routine childhood vaccination
Schedule • number of doses • timing • booster	 at least 3 IM as DTP 6/52, 10/52 & 14/52 of age DTP at 18/12 to 4 years of age
Contraindications & Precautions	Hypersensitivity reaction to previous dose or any constituent is a contraindication
Reactions/Side-effects	Mild local reactions common
Additional notes	Can also be DTP-HepB-Hib

Pneumococcus		
Vaccine type	7-valent conjugate Polysaccharide	
Indications for use	High risk, e.g. sickle cell, immunosuppresion, > 65 years of age, CRF, HIV, asplenia, chronic disease	
Schedule • number of doses • timing • booster	Conjugate 1–4 IM none	Polysaccharide 1 SC or IM consider 5-yearly
Contraindications & Precautions	Hypersensitivity reaction to previous dose is a contraindication	
Reactions/Side-effects	Mild local reaction Fever (conjugate)	
Additional notes	Polysaccharide not effective in < 2 years of age, little evidence of efficacy of polysaccharide	

Polio	
Vaccine type	Live attenuated oral polio vaccine (OPV, Sabin) Killed IM (IPV, Salk)
Indications for use	EPI Routine infant vaccination
Schedule • number of doses • timing • booster	 4 0, 6/52, 10/52 & 14/52 of age for endemic areas 10-yearly
Contraindications & Precautions	Children with rare immune deficiencies should have IPV not OPV
Reactions/Side-effects	Very rarely vaccine-associated polio (from OPV)
Additional notes	Both vaccines contain types 1, 2 & 3 OPV recommended due to low cost, ease of administration, superiority in intestinal immunity and potential for 2° immunity for contacts

Rabies	
Vaccine type	Inactivated virus (grown in cell culture)
Indications for use	Pre-exposure for at-risk populations, e.g. travellers, vets As part of post-exposure treatment
Schedule • number of doses • timing • booster	 3 IM or ID 0, day 7 & day 8–35 2–3-yearly depending on risk
Contraindications & Precautions	Hypersensitivity reaction to previous dose is a contraindication
Reactions/Side-effects	Mild local or systemic reaction Rarely neuroparalytic reaction reported
Additional notes	Animal vaccination in endemic areas Animal brain-derived vaccines associated with an increase in severe and fatal reactions

(*continued*)

Vaccines *continued*

Rubella

Vaccine type	Live attenuated virus
Indications for use	Routine of target immunisation Prime target women of childbearing age
Schedule • number of doses • timing • booster	 1 IM or SC 9–11/12 of age in highly endemic areas later for higher levels of control targeting of 15–40-year-olds none
Contraindications & Precautions	Hypersensitivity reaction to previous dose is a contraindication Pregnancy is also a contraindication
Reactions/Side-effects	Malaise, fever & rash 5–12/7 later Rarely encephalopathy and anaphylaxis
Additional notes	R, MR or MMR

Tetanus

Vaccine type	Formaldehyde-inactivated toxin (absorbed onto aluminium salts)
Indications for use	EPI Routine childhood vaccination Post-potential exposure, e.g. contaminated wounds
Schedule • number of doses • timing • booster	 3 IM or deep SC as DTP 6/52, 10/52 & 14/52 of age 10-yearly (or post-exposure if not up to date)
Contraindications & Precautions	Hypersensitivity reaction to previous dose is a contraindication
Reactions/Side-effects	Mild local or systemic reaction
Additional notes	T, DT or DTP Giving in pregnancy prevents neonatal tetanus Do not give at $<$ 10-yearly intervals due to increased risk of hypersensitivity

Tuberculosis	
Vaccine type	Live attenuated *M. bovis* (BCG)
Indications for use	EPI Routine infant/childhood immunisation
Schedule • number of doses • timing • booster	 1 ID at or as soon as possible after birth school age acceptable in low risk none
Contraindications & Precautions	Symptomatic HIV infection is a contraindication Pregnancy and generalised septic skin conditions are also contraindications
Reactions/Side-effects	Small swelling at 2–6/52 may → benign ulcer → heals at 6–12/52 Local abscess, regional lymphadenitis Rarely distant spread to give osteomyelitis of disseminated disease
Additional notes	Correct ID administration essential Except in infants leave > 3/52 between BCG and any other live vaccine Except in neonates, perform kind test prior to vaccination Shown to offer some protection against leprosy

Typhoid		
Vaccine type	Purified capsular polysaccharide Vi Live attenuated Ty21a (Killed whole cell – not recommended)	
Indications for use	High-risk areas and populations Travellers	
Schedule • number of doses • timing • booster	ViCPS 1 IM 3-yearly	Ty21a 3 oral as liquid or enteric coated capsules 0, day 2 & day 4 every 6/12
Contraindications & Precautions		
Reactions/Side-effects	Local reaction More serious with killed whole cell	

(*continued*)

APPENDICES

Vaccines *continued*

Typhoid

Additional notes	ViCPS < 2 years of age does not give long-lasting protection Stop antimalarials and antibiotics for 1/52 before and after Ty21a

Yellow fever

Vaccine type	Live attenuated virus
Indications for use	EPI Routine vaccination in endemic countries Travellers
Schedule • number of doses • timing • booster	 1 SC 9/12 of age, with measles in endemic areas As needed for others 10-yearly
Contraindications & Precautions	Hypersensitivity reaction to previous dose is a contraindication Pregnancy, symptomatic HIV infection, egg allergy and immune deficiency are all contraindications
Reactions/Side-effects	Rarely encephalitis in very young Hepatic failure Hypersensitivity to egg
Additional notes	Avoid in < 6/12 of age, need WHO approved centre to provide certification for travel across borders

Vaccination schedules

WHO-recommended infant immunisation schedule

Age[1]	Vaccines	HepB[2]	
		A	B
Birth	BCG, OPV0	HB-1	
6/52	DTP1, OPV1	HB-2	HB-1
10/52	DTP2, OPV2		HB-2
14/52	DTP3, OPV3	HB-3	HB-3
9/12	Measles[3]+/− YF		

[1] Babies born prematurely should be vaccinated at the same times after birth as babies born at term.

[2] A is recommended where perinatal HepB transmission is common, e.g. S.E. Asia; B may be used in countries where perinatal transmission is less common, e.g. SSA.

[3] Where there is a high risk of mortality from measles among children less than 9/12, e.g. hospitalised, HIV-infected, refugee camps, measles vaccination should be carried out at 6/12 and 9/12.

WHO/UNICEF recommendations for the immunisation of HIV-infected children and women of childbearing age

Vaccine	Asymptomatic infection	Symptomatic infection	Optimal timing of immunisation
BCG	Yes[1]	No	Birth
DTP	Yes	Yes	6, 10 & 14/52
OPV[2]	Yes	Yes	0, 6, 10 & 14/52
Measles	Yes	Yes	6 & 9/12[3]
Hep B	Yes	Yes	As for uninfected
YF	Yes	No	
TT	Yes	Yes	5 doses

[1] If local risk of TB infection is low, BCG should be withheld from individuals known or suspected to be HIV- infected.

[2] IPV can be used as an alternative in symptomatic HIV-positive children.

[3] Because of the risk of severe early measles infection, HIV-positive infants should receive measles vaccine at 6/12 and as soon after 9/12 as possible.

UK immunisation schedule

1st year of life
BCG at birth if at risk
DTwP, Hib, Meningitis C and OPV at 2, 3 & 4/12

2nd year of life
MMR at 12–15/12
Hib at 13–48/12

Before school or nursery
DTaP single booster dose
OPV single booster dose
MMR single booster dose

Between 10 & 14 years
BCG if unvaccinated and tuberculin-negative

Before leaving school or before employment or before further education
DT single booster dose
OPV single booster dose

During adult life
Rubella, Diphtheria, Tetanus & Polio if not previously immunised
Polio & Tetanus boosters every 10 years

MALARIA PROPHYLAXIS

(1) Protection against bites
- (a) long sleeves and trousers
- (b) DEET sprays etc.
- (c) Permethrin-impregnated mosquito nets
- (d) vaporised insecticides and coils

(2) Chemoprophylaxis
- (a) when to start before departure:
  2–3/52 mefloquine
  1/52 others
  1–2/7 atovaquone
- (b) when to stop after leaving area
  1/52 atovaquone
  4/52 all others
- (c) long term (seek specialist advice)
  3/12 atovaquone (licensed for 4/52)
  2 years mefloquine (licensed for 1) or doxycycline
  5 years chloroquine and proguanil
- (d) for use in children and pregnancy, and contraindications seek specialist advice
- (e) specific agents
  - (i) chloroquine
    - use alone in areas where resistance is still low
    - use combined with proguanil in areas where resistance is high (although it may not give optimal protection)
  - (ii) mefloquine
    - use in areas of high chloroquine resistance

- contraindicated in those with a neuropsychiatric history
- inform patients of adverse reactions

(iii) proguanil
- usually used in combination with chloroquine
- with atorvaquone and can be used in areas of high chloroquine and mefloquine resistance

(iv) pyrimethamine
- do not use alone
- with dapsone and can be used in certain areas of high resistance

(v) quinine
- not suitable for prophylaxis

(vi) artemisinin derivatives
- not suitable for prophylaxis

(vii) tetracyclines
- doxycycline used in areas of widespread chloroquine and mefloquine resistance

(3) Seek early blood film or rapid diagnostic test for diagnosis OptiMal, ICT – Plasmodium LDH detection – falciparum and vivax, ParaSight-F – PfHRP2 – falciparum only

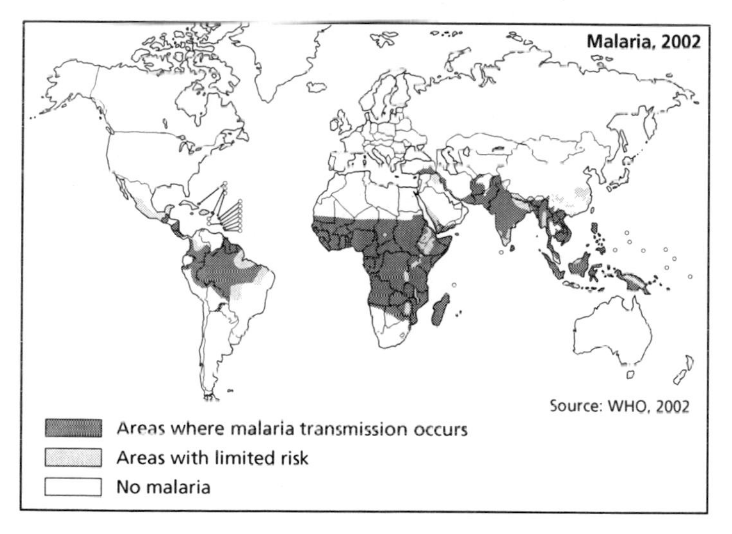

Worldwide distribution of Malaria in 2002. Source: WHO (2002)

MALARIA TREATMENT

If infective species is not known, or infection is mixed, treat with quinine or mefloquine or atovaquone.

(i) quinine/quinidine
- use for falciparum, unknown or mixed infection
- risk of hypoglycaemia, cardiac dysrythmias, hypotension

(ii) chloroquine
- no longer recommended for falciparum due to resistance
- can be used for benign malarias if species is known

(iii) artemisinin derivatives
- combined with lumefantrine – co-artemether

- artesunate, artemether, DHA all available orally, rectally, IM and IV injection

(iv) mefloquine
- use if infective species unknown or infection mixed
- do not use for treatment if used for prophylaxis
- do not use if severe disease (increased risk of post-malaria neurological syndrome)

(v) primaquine
- use for elimination of liver stages of vivax or ovale following chloroquine (radical cure)
- reduced dose if G6PD deficiency (causes haemolysis)
- kills falciparum gametocytes

(vi) proguanil
- with atovaquone and can be used for acute uncomplicated falciparum

(vii) pyrimethamine
- with sulfadoxine and can be used with (or after) quinine for falciparum
- no longer effective in most parts of the world

(viii) doxycycline
- use as an adjunct to quinine in drug-resistant falciparum (can use clindamycin if pregnant or child)

CHOOSING A DRUG

(1) choose a suitable drug based on knowledge about the patient and the likely causative agent
(2) local drug policies may have an effect on your choice
(3) take samples for culture and sensitivity before administration if possible
(4) decide on the correct route and dosage (correcting for renal or hepatic failure)
(5) take side-effects into account
(6) check for history of hypersensitivity

ANTIBIOTIC GROUPS

1. Penicillins: benzylpenicillin, phenoxymethylpenicillin, flucloxacillin (methicillin), amoxicillin, ampicillin, co-amoxyclav (amoxicillin + clavulanic acid), piperacillin, ticarcillin, pivmecillinam hydrochloride
2. Cephalosporins, cephamycins and other beta-lactams: cefclor, cefadroxil, cefalexin, cefamandole, cefazolin, cefixime, cefotaxime, cefoxitin, cefpirome, cefpoxodime, cefprozil, cefradine, ceftazidime, ceftriaxone, cefuroxime, aztreonam, imipenem with cilastatin, meropenem
3. Tetracyclines: tetracycline, demeclocycline hydrochloride, doxycycline, lymecycline, minocycline, oxytetracycline
4. Aminoglycosides: gentamicin, amikacin, neomycin sulphate, netilmicin, tobramycin
5. Macrolides: erythromycin, azithromycin, clarithromycin
6. Clindamycin
7. Some other antibacterials: nitrofurantoin, chloramphenicol, sodium fusidate (fusidic acid), vancomycin, teicoplanin, linezolid, quinupristin with dalfopristin, colistin, mupirocin
8. Sulphonamides and trimethoprim: co-trimoxazole (trimethoprim + sulfamethoxazole (SMX-TMP)), sulfadiazine, sulfametopyrazine, trimethoprim
9. Antimycobacterials: capreomycin, cycloserine, ethambutol hydrochloride, isoniazid, pyrazinamide, rifabutin, rifampicin, streptomycin, dapsone, clofazimine
10. Metronidazole and tinidazole
11. Quinolones: nalidixic acid, ciprofloxacin, levofloxacin, norfloxacin, ofloxacin

ANTIBIOTIC THERAPY

Cardiovascular system

Native valve

	Choice 1	Comments
Streptococcal endocarditis	benzylpenicillin (or vancomycin* if allergic) + low-dose gentamicin	check species and sensitivities
Enterococcal endocarditis	amoxicillin† (or vancomycin* if allergic) + low-dose gentamicin quinupristin/dalfopristin or linezolid if resistant	check species and sensitivities
Staphylococcal endocarditis	flucloxacillin (or benzylpenicillin if organism sensitive or vancomycin if patient allergic or organism methicillin-resistant) + gentamicin (or fusidic acid)	oral ciprofloxacin and rifampicin have been used for tricuspid valve
HACEK	ceftriaxone	
Bartonella	doxycycline or erythromycin	blood culture rarely positive

* Can use teicoplanin instead of vancomycin if once-a-day treatment preferred.
† Can use ampicillin instead of amoxicillin.

Respiratory tract

	Choice 1	Choice 2	Comments
Haemophilus epiglottitis	cefotaxime	chloramphenicol	give IV
Exacerbations of chronic bronchitis	amoxicillin	tetracycline (or erythromycin)	
Uncomplicated community-acquired pneumonia	amoxicillin (or erythromycin if allergic)	benzylpenicillin	cefuroxime if staph suspected + erythromycin if atypical suspected

Severe community-acquired pneumonia	cefuroxime (or cefotaxime) + erythromycin		
Atypical pathogen pneumonia	erythromycin	tetracycline alternative for chlamydia and mycoplasma	legionella may need rifampicin
Hospital-acquired pneumonia	broad-spectrum 3rd generation cephalosporin	antipseudomonal penicillin	+ aminoglycoside in severe disease

Gastrointestinal tract

	Choice 1	Choice 2	Choice 3	Comments
Gastroenteritis	not indicated			
Campylobacter	ciprofloxacin	erythromycin		wide variation in resistance to quinolones
Salmonellosis	ciprofloxacin	azithromycin	ceftriaxone	check sensitivity especially to nalidixic acid as a marker of reduced quinolone sensitivity
Shigellosis	ciprofloxacin	SMX-TMP		
Typhoid	ciprofloxacin	cefotaxime	chloramphenicol	
Antibiotic-associated colitis	oral metronidazole	oral vancomycin		
Biliary tract infection	cephalosporin	gentamicin		
Peritonitis	cephalosporin/ gentamicin + metronidazole/ clindamycin			
Peritoneal dialysis–associated peritonitis	vancomycin IP + gentamicin / cephalosporin +ciprofloxacin PO			can use teicoplanin instead of vancomycin

Urinary tract

	Choice 1	Choice 2	Choice 3	Comments
Acute pyelonephritis	broad-spectrum cephalosporin	quinolone		
Acute prostatitis	quinolone	trimethoprim		treat for 4 weeks
Lower UTI	trimethoprim	amoxicillin	nitrofurantoin or cephalosporin	can use ampicillin instead of amoxicillin

Genitourinary tract

	Choice 1	Choice 2	Choice 3	Comments
Syphilis	procaine benzylpenicillin	doxycycline	erythromycin	penicillin desensitisation if penicillin allergic with neurological disease
Uncomplicated gonorrhoea	ciprofloxacin	ofloxacin	cefotaxime	
Uncomplicated chlamydia, non-gonococcal urethritis and non-specific infection	doxycycline	azithromycin		
PID	ofloxacin	metronidazole		

Central nervous system

	Choice 1	Choice 2	Choice 3	Comments
Initial 'blind' therapy	ceftriaxone	cefotaxime	chloramphenicol	dexamethasone shown to be effective when given early
Meningococcal meningitis	ceftriaxone	cefotaxime		
Pneumococcal meningitis	ceftriaxone +vancomycin +/− rifampicin			

Haemophilus meningitis	cefotaxime	
Listeria meningitis	amoxicillin + gentamicin	can use ampicillin instead of amoxicillin

Eyes

	Choice 1	Choice 2	Comments
Purulent conjunctivitis	chloramphenicol eye drops	gentamicin eye drops	

ENT

	Choice 1	Choice 2	Choice 3	Comments
Dental infections	phenoxymethyl penicillin (or amoxicillin)	erythromycin	metronidazole	
Sinusitis	amoxicillin	doxycycline	erythromycin	
Otitis externa	flucloxacillin			
Otitis media	amoxicillin (or erythromycin if allergic)			can use ampicillin instead of amoxicillin
Throat infections	phenoxymethyl penicillin (or erythromycin if allergic)	cephalosporin		

Skin & soft tissue

	Choice 1	Choice 2	Comments
Impetigo local widespread	fusidic acid flucloxacillin	mupirocin erythromycin	topical oral
Erysipelas	phenoxymethyl penicillin		if staph + flucloxacillin
Cellulitis	phenoxymethyl penicillin + flucloxacillin (erythromycin alone if allergic)	co-amoxiclav	
Animal bites	co-amoxiclav		

Bones & joints

	Choice 1	Choice 2	Comments
Osteomyelitis and septic arthritis	clindamycin	flucloxacillin + fusidic acid	
Haemophilus	amoxicillin	cefuroxime	can use ampicillin instead of amoxicillin

Blood

	Choice 1	Choice 2	Choice 3	Comments
Septicaemia – initial 'blind'				choice depends on local resistance and clinical presentation
Community-acquired	aminoglycoside + broad-spectrum penicillin	broad-spectrum cephalosporin		
Hospital-acquired	+ broad-spectrum anti-pseudomonal penicillin	meropenem	imipenem with cilastin	
Septicaemia related to vascular catheter	flucloxacillin	broad-spectrum cephalosporin		
Meningococcal septicaemia	benzylpenicillin	cefotaxime		

ANTIBIOTIC PROPHYLAXIS

Surgical procedures

System/procedure	Choices
Cardiovascular (aorta repair, leg procedures involving groin incision, insertion of prosthesis/foreign body, lower extremity amputation, cardiac surgery, maybe permanent pacemakers)	2nd generation cephalosporin (e.g. cefuroxime) or vancomycin
GI	
gastroduodenal	2nd generation cephalosporin (e.g. cefuroxime)
ERCP (if obstruction)	ciprofloxacin or piperacillin
colorectal	
• elective	neomycin + erythromycin
• emergency	2nd generation cephalosporin (e.g. cefuroxime) + metronidazole
• ruptured viscus	2nd generation cephalosporin (e.g. cefuroxime), or clindamicin + gentamicin

Head & neck	cefazolin, or clindamycin +/– gentamicin
Neurosurgical	
• clean, no implant	2nd generation cephalosporin (e.g. cefuroxime) or vancomycin
• clean, contaminated	clindamycin, co-amoxiclav, or cefuroxime + metronidazole
• CSF-shunt	vancomycin + gentamicin
Obstetrics/gynaecology	
• Caesarian section or premature rupture of membranes	2nd generation cephalosporin (e.g. cefuroxime)
• hysterectomy	2nd generation cephalosporin (e.g. cefuroxime)
• termination	1st trimester: penicillin G or doxycycline 2nd trimester: 2nd generation cephalosporin (e.g. cefuroxime)
Orthopaedics	
• hip replacement/spinal fusion	2nd generation cephalosporin (e.g. cefuroxime) or vancomycin
• other joint replacement	2nd generation cephalosporin (e.g. cefuroxime) or vancomycin
• open reduction with internal fixation	ceftriaxone
Urology	
• pre-operative bacteriuria	2nd generation cephalosporin (e.g. cefuroxime) then nitrofurantoin
• transrectal prostate biopsy	ciprofloxacin
Others	
• peritoneal dialysis catheter placement	vancomycin
• traumatic wound (not bites)	2nd generation cephalosporin (e.g. cefuroxime) or ceftriaxone

Prophylaxis recommended for endocarditis

(1) high risk
- (a) prosthetic valve
- (b) previous bacterial endocarditis
- (c) complex congenital cyanotic heart disease
- (d) surgically constructed systemic pulmonary shunts/conduits

(2) moderate risk
- (a) most other congenital cardiac malformations
- (b) acquired valvular dysfunction
- (c) hypertrophic cardiomyopathy
- (d) mitral valve prolapse with regurgitation +/– thickened leaflets

(3) low risk
- (a) isolated secundum atrial septal defect
- (b) surgical repair of atrial or ventricular septal defect, or patent ductus arteriosus (without residua beyond 6/12)

(c) previous coronary artery bypass graft
(d) mitral valve prolapse without regurgitation
(e) physiological, functional or innocent heart murmurs
(f) previous Kawasaki disease without valvular dysfunction
(g) previous RhF without valvular dysfunction
(h) cardiac pacemakers & implanted defibrillators

Prophylaxis recommendations for the following procedures

System	Recommended	Not recommended
Dental	extractions; periodontal procedures including surgery, scaling, & root planning, probing & recall maintenance; dental implants; reimplantation of avulsed teeth; root canal instrumentation & surgery beyond the apex; subgingival placement of antibiotic strips or fibres; initial placement of orthodontic bands (but not brackets); intraligamentary local anaesthetic injections; prophylactic cleaning of teeth or implants where bleeding is anticipated	restorative dentistry; local anaesthetic injections (non-ligamentary); placement of rubber dams; postoperative sutures removal; placement of removable or orthodontic appliances; taking of oral impressions; fluoride treatments; taking of oral radiographs; orthodontic appliance adjustment; shedding of primary teeth
Respiratory/ENT	tonsillectomy; adenoidectomy; operations involving respiratory mucosa; rigid bronchoscopy	endotracheal intubation; flexible bronchoscopy; tympanotomy tube insertion
Gastrointestinal	sclerotherapy for oesophageal varices; oesophageal stricture dilatation; ERCP with biliary obstruction; biliary tract surgery; operations involving intestinal mucosa	transoesophageal echocardiography; endoscopy
Genitourinary	prostatic surgery; cystoscopy; urethral dilatation	vaginal hysterectomy; vaginal delivery; Caesarean section; in uninfected tissue – dilatation & curettage, termination, sterilisation, insertion or removal of IUCDs
Other		cardiac catheterisation; implanted pacemakers & defibrillators; coronary stents; incision biopsy of surgically scrubbed skin; circumcision

Regimens

	Dental, oral, respiratory or oesophageal procedures	Genitourinary or gastrointestinal (not oesophageal) procedures
High risk	amoxicillin	ampicillin + gentamicin
High risk, unable to take orals	ampicillin	ampicillin + gentamicin
High risk, allergic	clindamycin, cefalexin, cefadroxil, azithromycin or clarithromycin	vancomycin + gentamicin
High risk, allergic, unable to take orals	clindamycin or cefazolin	vancomycin + gentamicin
Moderate risk	amoxicillin	amoxicillin or ampicillin
Moderate risk, unable to take orals	ampicillin	amoxicillin or ampicillin
Moderate risk, allergic	clindamycin, cefalexin, cefadroxil, azithromycin or clarithromycin	vancomycin
Moderate risk, allergic, unable to take orals	clindamycin or cefazolin	vancomycin

Group B Streptococcus in labour

Vaginal colonisation with group B streptococci (GBS) is associated with ↑ maternal infectious complications and neonatal sepsis.
Treat during labour with penicillin G or ampicillin (cefazolin, erythromycin or clindamycin if allergic) if

(1) positive cultures from swabs taken at 35–37 weeks gestation
(2) risk factor(s)
  (a) previous delivery of infant with invasive GBS
  (b) GBS bacteriuria during pregnancy
  (c) delivery @ $<$ 37 weeks
  (d) duration of rupture of membranes $\geq$ 18/24
  (e) intrapartum temperature $\geq$ 38°C

If pre-term or premature rupture of membranes use ampicillin + erythromycin for 2/7 followed by amoxicillin + erythromycin.

NEEDLESTICK POST-EXPOSURE PROPHYLAXIS

Risks of seroconversion as a result of a needlestick injury from a +ve source:
HIV 0.3% HBV 30% HCV 6%
General management:
(1) wash wound (do not squeeze)

(2) report incident
(3) assess risk by
  (a) characterising exposure
  (b) determining/evaluating source of exposure by medical history & testing for HIV, HBV & HCV
  (c) evaluating & testing exposed person

HIV

Exposure type	HIV +ve (low viral load or asymptomatic)	HIV +ve (AIDS, high viral load, symptomatic, seroconversion)	Source of unknown HIV status, e.g. dead	Unknown source, e.g. sharps bin	HIV −ve
e.g. solid needle, superficial injury	basic 2-drug PEP recommended	expanded 3-drug PEP recommended	consider basic 2-drug PEP for source with risk factors	consider basic 2-drug PEP where exposure to HIV likely	No PEP
e.g. large-bore hollow needle, deep puncture, visible blood on device, needle used in artery/vein	expanded 3-drug PEP recommended	as above	as above	as above	No PEP

If PEP is offered and taken and the source is later determined to be HIV −ve, PEP should be stopped. Decisions where PEP can be considered should be based on a discussion between the exposed person and the treating doctor.
2-drug PEP: zidovudine + lamivudine
3-drug PEP: zidovudine + lamivudine + indinavir or efavirenz

HBV

Exposed person	HBVsAg^{+}	Source HBVsAg^{-}	Unknown
Unvaccinated	HBIg + vaccination (full)	vaccination (full)	vaccination (full) & HBVsAg on source
Vaccinated (Ab status unknown)	test Ab levels of exposed $\geq$ 10 MIU/ml, none $<$ 10 MIU/ml, HBIg + single dose of vaccine	none	test Ab levels of exposed $\geq$ 10 MIU/ml, none $<$ 10 MIU/ml, HBIg + single dose of vaccine

Accelerated hepatitis B vaccination course available.

HCV

No recommended PEP.

BOOKS

Cook, G. C. (Ed.) *Manson's Tropical Diseases*, 20th edn. W. B. Saunders: London, 1998.

Eddleston, M. & Pierini, S. *Oxford Handbook of Tropical Medicine*. OUP: Oxford, 1999.

Gilbert, D. N., Moellering, R. C. Jr & Sande M. A. *The Sanford Guide to Antimicrobial Therapy 2003*, 33rd edn. Jeb C. Sanford: USA, 2003.

Mandell, G. L., Bennett, J. E. & Dolin, R. (eds) *Principles and Practice of Infectious Diseases*, 5th edn. Churchill Livingstone: Edinburgh, 2000.

Ledingham, J. G. G. & Warrell, D. (Eds) *Concise Oxford Textbook of Medicine*. OUP: Oxford, 2000.

Mandal, B. K., Wilkins, E. G. L., Dunbar, E. M., *et al. Lecture Notes on Infectious Diseases*, 5th edn. Blackwell Science: Oxford, 1999.

Robbins, S. L., Cotran, R. S., Kumar, V., *et al. Robbins Pathologic Basis of Disease*, 6th edn. W. B. Saunders: London, 1999.

WEBSITES

www.aafp.org
www.americanheart.org
www.bnf.org
www.cdc.gov
www.doctors.net.uk
www.fitfortravel.scot.nhs.uk
www.liv.ac.uk/lstm
www.lshtm.ac.uk
www.mdconsult.com
www.netdoctor.co.uk
www.nhsdirect.nhs.uk
www.nih.gov
www.ukmi.nhs.uk
www.wellcome.ac.uk
www.who.int

TELEPHONE HELPLINES

Hospital for Tropical Diseases Travel Healthline 09061 337 733 (£0.50 per minute)

Recorded Advice for Travellers 09065 508 908 (£1 per minute)

Lightning Source UK Ltd.
Milton Keynes UK
UKOW04f1246230614

233899UK00001B/3/P